SVEN GEHRMANN

GENERAL FRESHWATER-CRUSTACEAN PRACTICE

Fresh- & Brackish-Water Crayfish, Shrimps, Prawns, Crabs, Hermits & Horseshoe Crabs

LOBBY FOR OUR RECENT NATURE

PROLOGUE

The first crayfish species I kept in a cold-water aquarium was the crayfish ***Orconectes limosus***, introduced from North America. At that time, in the mid-1980s, this species was, as a precaution, hardly mentioned at all in aquarium literature. The only reference I found was in Hans Frey's work "Aquarienpraxis kurzgefasst". There was a short section on the care of this North American crayfish, consisting of maybe 5-6 sentences and a black and white line drawing. At that time, nothing was published about crayfish pestilence and the problems associated with it, purely as a precaution. Tropical crayfish were rarely sold in the aquarium trade at that time, and then it was mostly the **Red Swamp Crayfish *Procambarus clarkii***, whose breeding was celebrated as a sensation in the aquarium magazines of the time! During the rest of my school years from 1985 to 1988, I was a member of a biology club that not only had many a water fight in the school's climate chamber during breaks, but also always acquired various interesting animals. During this time I procured crayfish from the edible crayfish trade for the first time, and managed to find **Swamp** and **Signal Crayfish**. During this time I also had my first experience with the crayfish pestilence, which killed all **Swamp Crayfish**. I had bought from "Nordsee" one by one. At about the same time I was looking for crayfish in a nearby gravel pond and found them rather by chance. Apart from the crayfish, I also caught the **Freshwater Shrimp *Atyaephyra desmarestii*** as a by-catch, which I could only identify with the help of the GDR reference book by Stresemann. A remarkable achievement, considering that at that time there was neither popular scientific literature on these animals nor the aid of the internet. This book thus contains my lifeblood, which has flowed over many years for these animals. You may forgive me for my strong language when I recommend cooking them in the chapter about the **North American Crayfish**, but this species is the first one I caught and kept in the aquarium myself. Since I sometimes also preserve animals and have occasionally dealt with their interior, please forgive me that I do not consider crayfish to be cuddly animals, nor do I intend to trivialise or humanise them. Rather, my intention with this work is to introduce the crustaceans of freshwater. I also want to contribute to creating a new awareness for the conservation of endemic species and their habitats. And finally, I would like to challenge the many aquarium enthusiasts in particular to deal responsibly with these versatile and interesting animals. If I were to advocate for anything, it would always be uncompromisingly for the cause of crustaceans

and the conservation of their habitats. Unfortunately, it is a sad fact, that every day about 250 species disappear irretrievably from our planet due to anthropogenic influences. Also, many researchers and climate experts now assume that climate change will mean certain extinction for about 30-40% of the animal and plant species known today. What never ceases to amaze me is the indifference and ignorance of most people towards these well-known hypotheses and facts. But what can we do ourselves to slow down the decline at least a little and to preserve the (still) existing biological diversity for our children? Well, this path may seem uncomfortable, but it starts with each and every one of us. In principle, it means renunciation. Renunciation of unnecessary consumption as well as renunciation of unnecessary luxury and unnecessary transport. Let me give you an example: The overproduction of meat in Germany has become increasingly grotesque in recent decades. In the meantime, grain has to be imported from countries such as Paraguay so that we can feed our own cattle, because in Germany it is no longer possible to produce such quantities of grain... This has dramatic consequences for the poor small farmers in Paraguay, who are simply driven off their fields to increase soy production and are also poisoned with pesticides. But the consequences of this madness are also becoming visible in Germany, because where fattening farms dispose of their manure, the nitrate levels in the drinking water are rising alarmingly, in some places to 100 milligrams of nitrate and more! Although I am not a pure vegetarian either, I am of the opinion that one can reduce one's meat consumption in order to counteract such circumstances. Even refraining from consuming pork can make a big difference, as the pig industry is unfortunately one of the most destructive forms of our agricultural economy. Did you know, for example, that about one third of the pigs produced do not even reach the market, but are "disposed of" in a completely senseless way? Another example of the most serious environmental destruction is the automobile. The production of just one car consumes up to 400,000(!) litres of drinking water, not to mention the electrical energy. And this is produced with harmful brown coal... Is there anything surprising about that? If we all don't learn to do without in the foreseeable future, we will soon feel the consequences of our unholy actions. For example, in the heatwave summer of 2018, river levels dropped dramatically, making even large bodies of water like the Main and Elbe partially unnavigable. From the North Sea coast to Saxony, the inland dried up... Therefore, it is a heartfelt request to my readers to seriously think about how to find good solutions on a small scale. I am convinced that not only the crustaceans of our inland waters will thank us for this. But also our children's generation. So, let's just be the miracle ourselves! Let's give up cheap flights, unnecessary car journeys, cheap rotten meat and disposable plastic products.

Sven Gehrmann, spring 2021.

CONTENT - INTRODUCTION TO A COMPLEX TOPIC

Crayfish from Australia

Crayfish from Europe

Crayfish from America

Infraorder *Caridea* – True Shrimps

Superfamily *Palaemonoidea* – Glass-Prawns

Infraorder Brachyura – True Crabs

Infraorder Anomura – Squat Lobsters, Porcelain Crabs & Hermit Crabs

Class Merostomata – Horseshoe Crabs (“Feeler-Less”)

GENERAL SECTION

A nerite rides around on this male Australian Red Claw....

Excerpt from the systematics of the *Arthropoda*

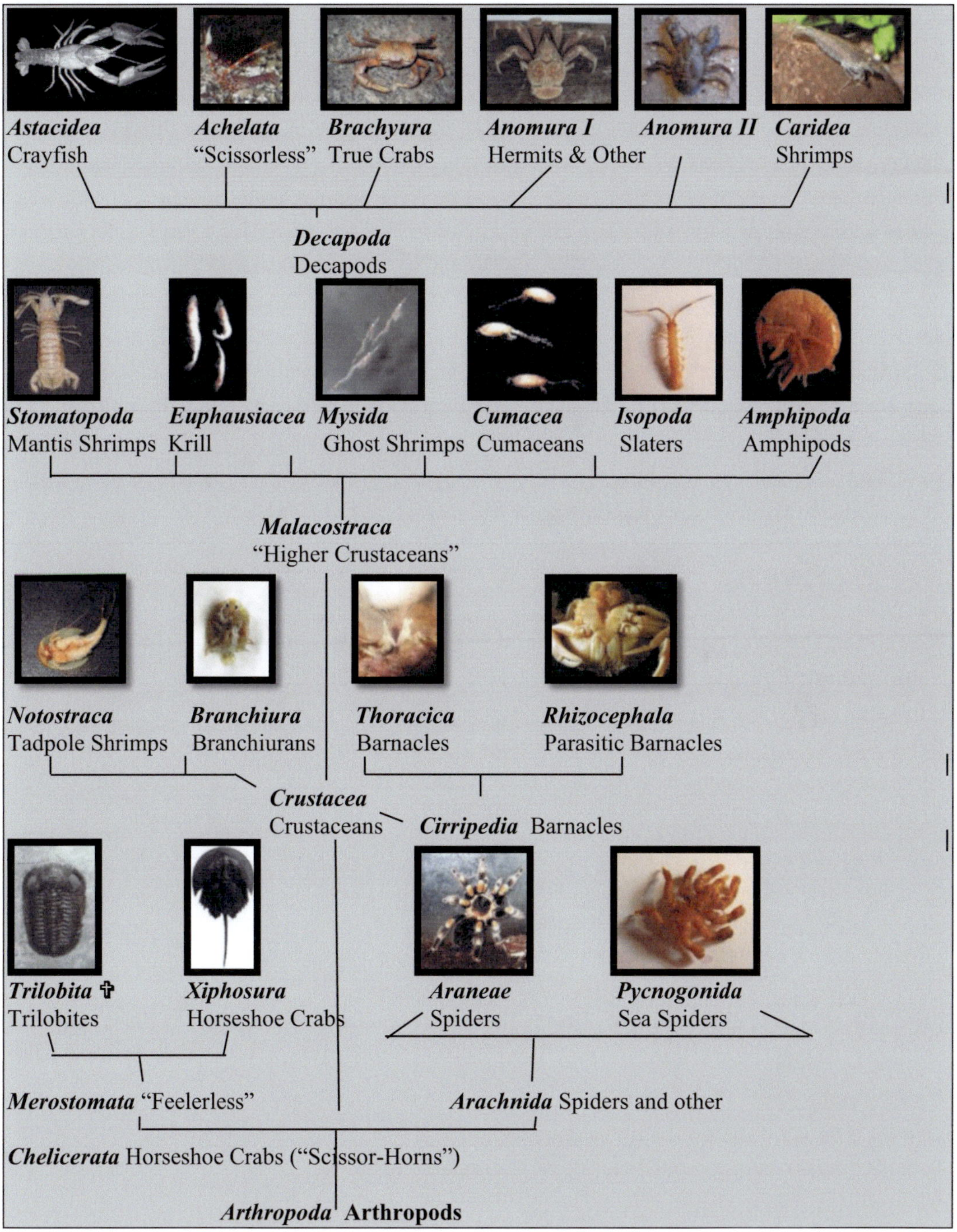

Some remarks on the meaning and use of systematics

When the ***Systema naturae*** of the Swedish scientist Carl Linne`(Latinised = "LINNAEUS") was introduced in 1758, a kind of "gold rush" will have broken out among scholars at that time. With the help of Carl Linne's guidelines on the system and determination of species, everyone wanted to make a name for themselves as quickly as possible by describing as many different species as possible. The scientists of the time were also vain fellows or were under pressure to obtain research funds from princes and kings, so they naturally wanted to show achievements. In addition, communication routes were long and international exchange was difficult in a Europe that consisted of many small states and principalities with many customs, borders and the like. Due to these circumstances, many animals were described twice and three times; some, such as the actinia of the genus ***Urticina***, were even described 42 times as a different species due to the colour variability of the species, although they were only 3 or 4 different species. I too have misidentified quite a few animals over the years, especially in the time when there was no internet and some genus and species revisions were being carried out. But how was the Carl Linne` system actually meant? Let us take a look at how a taxonomist would classify a **European Noble Crayfish**:

Domain: ***Eukaryota*** (Living being with cell nucleus and cell membrane) Whittaker & Margulis, 1978 – eukaryotes.
Kingdom: ***Animalia*** (Animals) Linnaeus, 1758 - Linnaeus, 1758 – animals.
Subkingdom: ***Bilateria*** (Bilateral animals) (Hatschek, 1888) Cavalier-Smith, 1983 - (Hatschek, 1888) Cavalier-Smith, 1983 – two sided animals.
Branch: ***Protostomia*** (protostomes) Grobben, 1908 - Grobben, 1908 – protostomes.
Infrakingdom: ***Ecdysozoa*** (ecdysozoans) Aguinaldo et al., 1997 ex Cavalier-Smith, 1998 - Aguinaldo et al., 1997 ex Cavalier-Smith, 1998 – animals, which must slough.
Phylum(Tribe): ***Arthropoda*** (Arthropods) Latreille, 1829 - Latreille, 1829 – arthropods.
Subphylum(Subtribe): ***Crustacea*** (Krebstiere) Brünnich, 1772 – Crustaceans.
Class: ***Malacostraca*** (Higher Crustaceans) Latreille, 1802.
Superorder: ***Eucarida*** (Long tailed Crayfish) Calman, 1904.
Order: ***Decapoda*** (Decapods) Latreille, 1802 – Decapods.
Infraorder: ***Astacidea*** (Crayfish and Chelate Lobsters) Latreille, 1802 - Crayfish and Chelate Lobsters.
Family: ***Astacidae*** (True Crayfish).
Genus: ***Astacus*** (Crayfish) (Linnaeus, 1758).
Specific name: ***astacus*** (Noble Crayfish) Linnaeus, 1758.

Very many scientists were involved in creating this taxon-tree over many ages…

Scientific name (Genus and species): ***Astacus astacus,*** **Linnaeus, 1758.**

Due to international rules, the systematics of animals is mostly presented in English and Latin or Greece. Every living being on our planet can be assigned its place in this system. The best way to imagine this system is as a large shelf with many niches that have certain designations and in which the corresponding creatures can be sorted. However, Carl Linne's original system has been expanded and improved many times in the meantime, as more and more animals and plants have been discovered, that cannot be classified into simple categories, as they often move systematically between different gradations or represent transitional forms. In the example above, the scientist and year to which each stage is to be assigned was indicated behind it. This shows that the taxonomy of living organisms can still change and expand; therefore, this system must not be understood as a rigid static structure. Rather, the structure grows with the state of research, and it is conceivable that the example shown above will need twice as much space in a hundred years' time, viewed from top to bottom, because even more fine gradations have been added. I consider it particularly important to indicate the name of the first describer of a species and the corresponding date of the description after the Latin double name, because one can only assume a secure species if one has found these dates. Unfortunately, some species and/or genera are checked and revised again and again, so that one can never assume that the result of a determination will last forever. Therefore, if you want to maintain a professional collection, you should always check the topicality of the nomenclature at intervals of a few years with the help of up-to-date internet sites. For this purpose I particularly recommend the use of ZIPCODEZOO.COM. If possible, one should prefer official sites of the respective universities and always take into account that even the experts unfortunately do not agree on some things so quickly and in some cases even publish different data. In my excerpt from the Systematics of Arthropods, I myself have divided the **Anomura** into **Anomura I** and **Anomura II**, because I think it is likely, that a revision will still be made in this area. This is because Hermit Crabs on the one hand and porcelain and King Crabs on the other show very considerable morphological variations, so that one has to wonder here whether the classification of these animals in the same order is correct. It is my hope that molecular genetics will provide new results on this in the near future. And that taxonomists do not invent more taxa that complicate the systematics even more... Finally, it should be noted that the concept of species in biology is unfortunately not clearly defined at present. Molecular geneticists are arguing with morphologists and others; the outcome is uncertain. Is there any objective truth here at all? Probably everything is in flux at the moment! Only one thing is certain: it is not as simple as we were taught in school thirty years ago. We now know that species evolve from other species. And that in some cases even apparently different species can interbreed and produce fertile offspring. Until recently, this was considered an important criterion for species delimitation. It is possible that the importance of molecular genetic analyses is simply overestimated or even misinterpreted at the moment. Therefore, not everything that comes from this branch of research should immediately be considered objective and true. After all, the scientists of this faction could ultimately be just as addicted to profiling as their colleagues were in Carl Linnes` days. And unfortunately, it is also a deplorable shortcoming, that research work nowadays tends to be project- and goal-oriented. Because research also costs money, of course, and hardly anyone can work for free or for God`s wages. As a result, today's scientific world is not opinion-neutral or independent, which is why it often adopts the premises of its clients. This then distorts the results accordingly. So one should not be too faithful to science and better not expect value-neutral objectivity. On the other hand, you can also become a researcher yourself by writing down your observations and communicating them to others. In this way, even hobby researchers can often make a valuable contribution that is above all not unobjective due to commercial requirements. (Some hobbyists have even described new species). Become an unbiased nature researcher, too!

Keeping crayfish in aquaria

Let's be realistic: we will never succeed in recreating nature's model 100 per cent in our aquaria. All we can do is to make life in the aquarium as pleasant as possible for our animals, so that they feel completely at ease and also reproduce here. In doing so, it may be, that what is not pleasing to the eye is ideal for the animals and vice versa. On the one hand, crayfish do not like sterile aquaria that always look clean and tidy, as if a cleaning fairy had been at work only yesterday, but on the other hand, the question of water quality should not be neglected. I would therefore describe optimal crayfish tanks as being characterised on the one hand by consistently good filtration and regular water changes, but on the other hand also by a wealth of structures, cover and retreat possibilities for the animals and by well-kept detritus accumulations in certain places. My crayfish aquaria are usually set up according to the same basic idea, which I would like to briefly introduce here. But let's start with the substrate: I prefer layers of fine sand, whereby the layer height of the sand can be up to 10 centimetres or more. In this concept, the substrate is also used as a filter. With a substrate placed in this way, three layers form after a short time, which are important for water purification. The uppermost layer up to a depth of about 5 centimetres can be described as an oxic layer, in which there is still a relatively high oxygen content. - This is followed by a suboxic layer, which runs from about 5 centimetres to 8 centimetres deep and already contains considerably less oxygen. Below this is an anoxic layer, in which numerous anaerobic bacteria live, which utilise the metabolic waste products of other organisms. This layer in particular acts like a natural sewage treatment plant and breaks down nitrate. Blue-black bacterial discolouration often occurs in this layer. Especially in this lowest anaerobic zone, nitrate is split into pure nitrogen and pure oxygen by the anaerobic bacteria groups, so that an enormous nitrate decomposition takes place. Gas bubbles form in the substrate, which leave the aquarium after a while via the water surface. The older this substrate gets, the higher its biological efficiency. The main advantage over other substrate materials, such as gravel, is, that due to the fineness of the sand, it is relatively difficult for organic waste to penetrate these anoxic layers, so that no foci of putrefaction with sulphates (sulphur compounds) can form here. Another advantage of sand is, that it is a very inexpensive substrate that can be partially vacuumed out and renewed at any time. I have found commercially available bird sand, from which I rinsed out the aniseed oil before using it at home, to be a good choice. Parrot sand, on the other hand, should not be used, because it contains caustic lime, which unfortunately makes it unusable for aquarium purposes. Once you have created such a biologically functional substrate, you no longer need any elaborate filter systems, because the substrate already performs this function. Nevertheless, the strongest possible water circulation should be ensured by one or more pumps. In my crayfish tanks, the external filter of the brand "EHEIM-Liberty" has proved particularly effective, as it brings a strong, adjustable current into the water and at the same time removes considerable amounts of organic waste from the aquarium through easy-to-clean filter cartridges, so that it can be taken out of the water cycle. Another function of the sandy bottom, that should not be underestimated, is, that the crayfish can also occupy themselves sensibly here - just like in nature - by digging. I am convinced, that some animals become aggressive in the aquarium, because they do not have these opportunities to occupy themselves. Therefore, one should tolerate the redecoration of the tanks by the crayfish within certain limits. All structures in a crayfish aquarium should be built in such a way that the crayfish cannot bring them down. If larger stones are to be used, they should be placed on polystyrene or Plexiglas sheets to prevent the bottom from cracking. This is especially true when keeping larger crayfish species, which can reach final sizes of 10 centimetres and more. I think planting is rather unsuitable for most crayfish tanks, as most crayfish regard plants as food. You can help yourself with foliage and plastic plants. Foliage is a staple food of most crayfish anyway and promotes their health, as beech foliage, for example, contains substances that act like antibiotics. And many crayfish species also like waterweed very much. These natural foods contain numerous vitamins and dietary fibres that significantly increase the vitality of the animals. Conversely, the absence of these nutrients can produce pigment deficiency and moulting problems. Therefore, you should always make every effort to provide your pets with the best possible nutrition.

Plastic plants are a real alternative for crayfish tanks with herbivorous crayfish species; you might want to attach them to the bottom of the aquarium with some silicone. By the way, silicone also hardens under water and can thus be added to the aquarium afterwards.

Artificial rock backwalls are very decorative, but must be secured with Plexiglas plates at the top, otherwise the crayfish can climb behind them. Here a tank with Apricot Crayfish *Cherax holthuisi* and Dwarf Rainbowfish *Melanotaenia praecox*.

This aquarium was stocked with Dwarf Rainbowfish and Apricot Crayfish. *Cryptocoryne wendtii* proved to be a grateful and durable plant.

Wooden roots can be very decorative and therefore I can recommend two types of roots here: Either you use roots that have been in another aquarium for a long time and are really sure, that they do not release any harmful substances such as nitrite into the water. Or you use roots made of the non-toxic Polyrin. Another option is to find your own natural dried wood. I only use pieces of wood, that are bone dry and faded and have been lying around for several years. No algae or fungi should grow on these pieces of wood. If you water such dried wood, it usually sinks after 1-2 days. After successful watering, perching plants such as **Java Moss** or **Java Fern** can be attached to such wooden branches with fishing line and allowed to grow. This gives a very nice visual effect to the aquarium and you have plants that are not too exposed to the vagaries of the crayfish. Stones can be added to almost any aquarium as decorative material without hesitation, if they do not contain any metallic or other harmful inclusions. If the tap water is already very hard, you should avoid using limestones, as they can unnecessarily increase the water hardness. If you lay stones on the sandy bottom, the crayfish have the opportunity to dig hiding places between the stones, just like in nature. Such crevices and cracks later provide numerous hiding places for the crayfish fry. Under no circumstances should the crayfish fry be reared in sterile breeding tanks, as the young animals absolutely need infusoria and other organisms, which they find in the detritus for their healthy development! Excessive hygiene in a crayfish aquarium is therefore more harmful than beneficial. On the one hand, the adults are disturbed by all the "piddling around" in the aquarium, and on the other hand, the fry cannot develop. It is very bad to expose egg-bearing females to stress in any form. This usually causes them to lose their brood! Larger water changes should be carried out depending on the amount of food given and the stocking density. Quantities of 50 - 80% of the aquarium water should be changed, because smaller changes hardly cause a short-term improvement of the water quality. If you have had deaths in a crayfish aquarium due to moulting problems, you should check the nitrate value and carry out a water change if necessary. Larger water changes should be carried out once or twice a month in heavily stocked aquaria and once a month in lightly stocked aquaria. Because of the crayfish pestilence, European, American or Australian crayfish species should not be kept together. Different species from one crayfish habitat, which even occur together in nature, should not be put together in the confines of an aquarium, as one species will usually crowd out the other. Such husbandry attempts are only possible and promising in exceptionally large tanks. However, it is possible to keep crayfish together with harmless shrimps, thick-shelled snails, or freshwater mussels. Large crayfish can be kept together very well with dwarf shrimps of the ***Caridina-*** and ***Neocaridina genera***. When socialising with fish of any kind, you should consider from the outset that you cannot treat fish diseases in a crayfish tank with medication, because this can be harmful to the crayfish. Therefore, I recommend at this point to use a UV device that kills the harmful germs. The UV device should always be used in a slightly oversized way so that it produces the optimal effect. This means, for example, that for a 100-litre aquarium you should use a device that is designed for 200 litres. The UV radiation is completely harmless for the animals, as it remains in the device and only disinfects the water. With this method, even a severe infection of the fish with ***Ichthyophtirius multifilis*** can be successfully contained. Veil fin breeds or bottom-oriented fish should better not be put together with crayfish. Crayfish species with long slender claws often tend to grab fish deftly, while species with rounded claw shapes usually do not chase fish. If you associate animals that do not naturally occur in the same habitat, there is no natural predator-prey scheme here, and keeping them together is usually successful. An example of this would be keeping Australian crayfish together with medium-sized African cichlids from Lake Malawi, with which I have had very good experiences. These should be just a few general care tips and inspirations from many years of experience with keeping and breeding a wide variety of crayfish species. Of course, different species have different husbandry requirements, but I hope that one or the other crayfish keeper or newcomer can benefit from some of my ideas and experiences.

When crayfish clean their eyes, it is always a funny sight. Perhaps this is also the attraction of taking a closer look at these animals.

Nutrition of the crayfish

This *Cherax boesemani* "Red Brick" is gnawing on a piece of dead wood.

In this topic, a fundamental distinction should first be made between keeping crayfish in a pond and keeping them in an aquarium, as the confined space of an aquarium brings other problems with it, than keeping them in a pond. In a pond crayfish usually already find everything they need to live. Which is why they only need to be fed separately in commercially operated aquacultures. Any kind of unsprayed(!) grain, (pre-cooked) carrots, cucumbers and lettuce, as well as feed granules are suitable for this. Crayfish in particular react to chemically contaminated food with unusual behaviour, panic reactions or even an attempt to leave the water. Again and again one encounters the ineradicable prejudice, that crayfish are fish eaters. This is not true in this form. Crayfish are omnivorous animals, that only eat fish, when they are sick, or if they cannot escape the crayfish in the cramped conditions of an aquarium. Most crayfish like to eat alder, willow and beech leaves, plant material, detritus, algae and small aquatic animals of all kinds. Some crayfish even eat dead wood, like the ***Cherax boesemani*** pictured above. Unfortunately, it is not known whether they eat the wood to absorb fibre, or whether they actually eat it as a complete food. It is important to make sure that the crayfish are fed as varied and also filling a diet as possible to avoid encroaching on moulting conspecifics. Excessive feeding of high-protein foods should be avoided, as some crayfish are otherwise prone to shock moulting. This means, that they moult too early and then die, because its new carapace has not yet developed far enough. For the crayfish I keep, the spirulina-containing herbivore food Tetra Pro Vegetable has proven to be the main food. In addition, I also feed black, white and red mosquito larvae as well as Mysis. I also occasionally feed frozen smelts to my cold-water crayfish. You should rarely feed crayfish with dead fish, as otherwise you could give them too much of a fishy taste, which could then cause the fish to fail in the aquarium.

Here a Yabby is eating a dead fish. But under no circumstances would it think of running after healthy fish on its own. That would be a waste of energy for him!

Especially the crayfish fry needs detritus for its development, as *Infusoria* and other nutrients live in it.

Ponds for crayfish

Ponds planted with water lilies keep away some American crayfish species, such as the Red Swamp Crayfish *Procambarus clarkii*, because they release substances into the water, which these animals do not like.

The Long-Clawed Swamp Crayfish *Pontastacus leptodactylus* can also be kept in the pond.

Due to the crayfish pestilence and faunal distortion problems, only native crayfish species should be kept in garden ponds, as the migration of crayfish can never be completely prevented. Ponds that are at least one metre deep are suitable for the successful keeping and breeding of crayfish, so that they cannot freeze in winter. The pond should be richly structured and offer wooden roots and hiding places under large stones at the bottom. Ideal is a plant stocking, which can also be used by the animals as food. Heavily muddy ponds are not suitable for the most crayfish, because they would constantly stick in the mud and will try to migrate overland due to avoid that. If you want to breed crayfish as efficiently as possible, it is better not to stock them with large fish species, as they will definitely decimate the broods. When stocking ponds with crayfish, one crayfish per metre of shore length was always given in the literature in the past, but the number of animals that can be used also depends on the size of the juveniles and the nature of the shore. Based on my experience, I think stocking with one- to two-summer juveniles per metre is most suitable, because larger animals tend to migrate from a pond over land more quickly. And because larger females do not carry as many eggs as smaller female crayfish. Membrane ponds are preferable for stocking with crayfish, because ponds that have been artificially constructed with mud walls, will almost certainly be perforated and drained by the crayfish. Although juvenile crayfish live in warm shore areas among algae and water plants, the adults like to dig tubes several decimetres long into the pond walls. However, there is nothing wrong with using crayfish in natural spring or seepage ponds, as they can hardly cause such damage here. Crayfish need beech, alder or willow leaves as additional food, which suggests planting such woody plants near the pond. Willows with their roots protruding into the pond from the pond bank look particularly beautiful and natural, but especially in hot summers, they can draw considerable amounts of water from the pond due to their evaporative properties. Crayfish thrive best in waters that are rich in nutrients and warm in summer. If they are introduced into nutrient-poor, cold streams, they may migrate, because they do not prefer such conditions. In breeding ponds perforated stones and clay bricks should be used as cover for the crayfish fry, and plastic pipes or clay drainage pipes up to about 8 centimetres in diameter are also suitable for adults. What are the advantages for the pond owner, when stocking with crayfish? Apart from the fact that the pond owner makes an active contribution to the conservation of endangered species, he is rewarded for his efforts in many ways. Crayfish keep the water clean by eating dead animals and leaves that have fallen into the water. In this way, the animals themselves make a significant contribution to water hygienics. At the same time, the pond owner can tell from the behaviour of his pond inhabitants whether everything is all right with the water. Since crayfish react very sensitively to disturbances in the water chemistry, they can indicate problems to the pond owner by their unusual behaviour.

Pond hygienics

In order to prevent the infection of a crayfish pond with crayfish pestilence spores, care should be taken to ensure that all equipment, nets, etc. used in the pond are never used in other waters. Under no circumstances should you fish in crayfish waters with bait made of crayfish meat, as this can easily introduce crayfish pestilence spores. Care should also be taken not to introduce crayfish spores from other waters with rubber boots. Rubber boots can be effectively disinfected with quicklime or acetic acid. When stocking a crayfish pond with fish from other waters, they should first be kept in quarantine, whereby keeping them with Ultra Violet treatment is optimal, as this kills all pathogens floating in the water. If you discover dead crayfish in your pond, they should be removed from the water in any case, as crayfish consume dead conspecifics and can thus become infected with various diseases, such as porcelain disease or burnt spot disease. If large numbers of crayfish die, the water should be tested and an animal possibly examined by a veterinarian to rule out an infestation with crayfish pestilence. If one wants to preserve a dead animal, it should be kept in a glass with 40% alcohol or in a 5% formalin solution. However, these preservations should not be exposed to the sun, because otherwise the crayfish will turn red and lose their natural colouring. The jars must be filled to the top with the preservation liquid, otherwise the essential components of the solution can evaporate and the preservation then spoils. As a final note on pond hygienics, water birds should be kept away from crayfish ponds, as they can easily introduce spores of crayfish pestilence from other waters. For this purpose, it is best to use plastic dummies of the corresponding bird species (mostly mallard and heron).

River Snails and mussels clean the water by filtering and eating detritus.

Keeping crustaceans in commercial facilities

In the food trade, crayfish from Iran are usually offered. In this case it was the Long-Clawed Swamp Crayfish *Pontastacus leptodactylus*. Very often there are also females with eggs. Whether these develop after acclimatisation in an aquarium depends on when the eggs were released. If the eggs were released before export, usually only very few young crayfish develop, as most eggs are damaged by the waterless transport and then later become fungal. They must never look yellowish, but brownish, greenish or dark. If you are lucky, you can sometimes get pretty bluish crayfish. With these, their eggs look bluish, too.

In the meantime, refrigeration technology has advanced so far, that it is no longer a problem to keep edible crustaceans alive in refrigerated containers for days and weeks. In addition, the procurement of synthetic sea salt is no longer a problem and the production of artificial seawater has become a simple process. Nowadays, synthetic sea salt can be dissolved in half an hour and the water - when it is clear and all the salt is evenly distributed - can be used immediately. Thus, even lobsters, langoustines and even the Arctic king crab and Norway lobster can now be kept alive professionally for longer periods of time for the gourmets. The same applies, of course, to crayfish and various mussels. When handling food animals, it is therefore mostly important to acclimatise the newly delivered animals properly and to prevent attacks of the crustaceans against each other by claw restraints or individual keeping. The handling of the individual species can be quite different. Food animals are not fed in order to avoid spoiling the water in the enclosure through uneaten food. In addition, dead animals must be meticulously collected for food safety reasons alone. In any case, keeping crustaceans alive guarantees the highest possible quality of the crustacean and protects the consumer from spoiled food and possible fish poisoning. In this respect, such facilities represent a valuable contribution to the healthy nutrition of consumers. I also think it is possible to make a contribution to the environmentally sound education of children, as they can learn, that it is animals we can touch, that we get food from, and not finished products from the freezer at the nearest supermarket. Finally, a facility of an importer of aquarium animals is shown here, in which the animals were housed in a species-appropriate and efficient manner with the simplest of means.

Live Purple Crayfish in the facility of a well-known German importer. Plastic tubes were used as hiding places, as this is the easiest way to catch the animals for sale. On the other hand, they still feel relatively comfortable because they have enough hiding places. In addition, some sand was brought in so that the animals can dig and so that aggression against each other is contained. In this facility, some crayfish even produced offspring!

Preservation of crustaceans with natural colours

It is best to use cake - or lobster forks to remove the flesh. From all parts of the crustaceans body, which contain a particularly large amount of tissue, this must first be removed as completely as possible. Care must be taken not to break the shell. Therefore, start by opening the carapace on the dorsal side of the crustacean at the seam between the ***Cephalothorax*** and the ***Abdomen***. Now the tissue can be carefully removed using a lobster fork or tweezers. Then carefully remove the claw joints and take the meat out of the claws. Since the claw legs also contain a lot of body tissue, one should cut off the claw segments completely and hollow out the claw legs as well. With medium-sized crustaceans you can save yourself this time-consuming procedure, as they do not contain much body tissue here. Then place the crustacean shell in about 40% alcohol or in a 2-3% formalin solution and leave it in this bath for about two weeks. If the alcohol or formalin content is too high, the preparation will inevitably turn red. Afterwards, the dishes are fixed on styrofoam plates to dry with pins in the desired position. However, the preparation must not be placed in the sun under any circumstances, otherwise the colour pigments of the crustacean will be destroyed and the preparation will turn red. Broken limbs can be glued back on with a hot glue gun or superglue. It is advisable, to put the dried specimen in an airtight display case for permanent storage, so that no parasites, such as meal moths or museum beetles, can settle in it. Ideal display cases are wooden wine boxes in which a cut-to-size glass pane is inserted instead of the wooden flap. Discarded drawers, that have been covered with an appropriately cut wooden or Plexiglas pane have also proved successful. If you only want to keep voucher specimens of crayfish, you can also store them permanently in 40% alcohol or 3% formalin, for which tightly closing screw-top jars of all kinds or preserving jars are particularly suitable. These jars should then also be labelled accordingly to keep track of the collection that has been created. The labels should be attached to the outside of the jar, so that they are clearly legible and should be covered with transparent adhesive film to protect the labels against leaking liquids and moisture. Such a collection offers the advantage of being able to carry out examinations and comparisons on the animals even years after they have died.

Ethics and crayfish

Unfortunately, due to the general invertebrate boom in aquaristics in recent years, the enthusiasm for exotic crayfish, which until then had been largely untapped, has been carried on very uncritically by the broad general public of aquarium enthusiasts. In some cases, commercial short-term profit interests were at the forefront of import efforts, and even today there are companies, that want and need to import new species for image reasons at almost any price. While I understand that there are also economic constraints and dependencies here, I would like to ask all true nature lovers at this point to refrain from the ambition of always wanting to acquire and breed newly imported species. Rather, I am of the opinion that new colour morphs of already imported species can also be sorted out through targeted breeding, so that new imports can be reduced to a minimum.

***Cherax spp.* "Orange Tip", shortly after import at the importer.**

It is unfortunately a very critical point that especially in the field of crayfish, species are still being imported that can become a danger to our native crayfish. The spread of these animals and their offspring can no longer be controlled. While the majority of these species cannot survive our cold winters in the wild, some species are more adaptable than thought, and not only survive our cold winters, but also reproduce in our waters. The ability of crayfish to migrate from farmed and garden ponds under wet conditions and reach other water bodies is also unfortunately usually underestimated. The ecological impacts to be expected as a result can be easily foreseen due to past introductions: On the one hand, our native species will become rare "exotics" due to dramatic population declines, and on the other hand, in some water systems, new introduced species will always displace the previously introduced species. The effects this can have on the entire ecosystem of a body of water are incalculable. The fact is, however, that tragedies take place here every day that Mother Nature can only weep over. Commercial interests of business enterprises then possibly finish off the last refuges of our autochthonous crayfish stocks. It would therefore be very desirable if nature-loving people would stock their garden pond biotopes exclusively with our endemic crayfish species, so that our once most common crayfish does not suffer the same fate as the giant auk or the North American migratory pigeon. A dry note in the biology textbook: Extinct. Therefore, at this point we should also contradict those, who claim that the spread of new species is an evolutionary process. This is unobjective and incorrect. Strictly speaking, this is an unnatural process of displacement, the effects of which can no longer be understood as evolutionary (new species are created), but only as devolutionary (species disappear completely). The effects of this decrease in biodiversity will increase dramatically in the future and ultimately affect humans themselves. At this point, it is my appeal to all nature lovers to do some deep soul-searching here. And consider, what they themselves can do to stop this trend. Nothing can justify the spread of neozoans/neobiota! Finally, I would like to point out, that handling freshly imported crayfish from overseas can also pose health risks, as the crayfish often serve as intermediate hosts for dangerous parasites that can also attack humans. The main ways to protect yourself from this are not to touch new imports with your bare hands, not to eat them raw(!) under any circumstances, and to be careful not to suck water from an aquarium with imported animals with your mouth. Infection with flukes from crustaceans and snails can cause severe damage to the liver, intestines and other vital organs and should be taken very seriously in any case. If left untreated, one can also die from such parasites, as some parasites can cause multiple organ failure after a while. Therefore, good hygienics is indispensable when importing from overseas. Carelessness can even be fatal!

Family *Parastacidae* – Crayfish from Australia

More than 100 different species of crayfish are endemic to Australia and New Guinea. These crayfish are usually classified under the collective term **"Australian crayfish"**, although strictly speaking this is not correct. They belong to the **Family *Parastacidae***, which is distributed in the southern hemisphere. Other representatives of this family are found on Madagascar and in South America. It is quite conceivable that a revision of the "Australian" crayfish species could sooner or later lead to the definition of new crayfish families, since some of the Australian species have already adapted very extremely to very specific habitats and have formed their own developmental lineages. There are even species that live more on land than in the water, and which have thus conquered a place that would otherwise have to be occupied by certain crab-species.

"True" Australian crayfishes

I would like to define as genuine Australian crayfish those species, that actually occur on the main Australian continent. These species are basically only available from the aquarium trade, because the Australians also trade them as edible crayfish. They are now farmed in aquacultures worldwide. These species include, for example, the **Yabby *Cherax destructor*** or the **Australian Red Claw *Cherax quadricarinatus***. All non-edible crayfish, including the small species that are of particular interest to aquarists, are strictly protected in Australia and are not allowed to leave the country. The Australians are very sensitive to the illegal export of animals and plants and punish so-called biopiracy very severely. For example, a woman who tried to take strictly protected snakes out of the country was fined 10,000 Australian dollars and sentenced to 10 years in prison. This example only serves to illustrate, that this is not a trivial offence, and that one should be very careful when dealing with this matter, especially in Australia. For these reasons, the number of "genuine" Australian crayfish species in the trade is severely limited. Nevertheless, there are still many possibilities for keeping and breeding these species for the committed aquarist, and we can expect to see some new and colourful breeding forms in the near future.

"False" Australian crayfishes

From a purely morphological point of view, these species often belong to the same genera as the "true" Australians, but they are not found on the Australian mainland. These species mostly come to us from New Guinea, especially Irian-Jaya, and are often very colourful animals. These include animals like the **Apricot Crayfish *Cherax holthuisi*** or the nice purple **Boesemann's Crayfish *Cherax boesemani***. We can expect some new discoveries and interesting imports from this region in the foreseeable future. Of course, I would not include the **Yabbys** bred in Israel or the USA among the "false" Australian crayfish, but all those species, that do not naturally occur on the main Australian continent.

Habitats of the Australian crayfish

The habitats of Australian crayfish are very diverse and cannot be described in more detail here due to lack of space. However, it should be noted, that there are Australian crayfish species, such as the **Koonac *Cherax preissii***, that require sea salt additions to the aquarium water to reproduce, because they naturally occur in brackish water habitats. Still others occur in flowing waters, while there are also species that occur in temporary waters. Some are even capable of surviving years of drought buried several metres deep in the mud of a lake. Then there are also some that live almost like insects in damp forests and are actually more land than water dwellers. Finally, it should be mentioned that there are also cave-dwellers among the Australian crayfish species that occur in habitats without plant growth. These are called troglodytes. A typical species of this group is the **Apricot Crayfish *Cherax holthuisi***.

General information about the care and acclimatisation of Australian crayfish

Australian crayfish are just as little immune to crayfish pestilence as European crayfish. Therefore, keeping them in the same water cycle with American species is out of the question. In principle, only one species per tank should be kept, as otherwise mating of crayfish of different species cannot be ruled out. Also, crayfish of different species usually crowd each other out, which then leads to losses. When acquiring newly imported species, they should be acclimatised in a quarantine tank where the crayfish can find sufficient hiding places. Well-watered foliage should not be missing, as the crayfish need this as important food. Lighting should be used very discreetly in such tanks or not be present at all, as freshly imported animals are very shy at first. The aquarium should be large enough so that the animals can get out of each other's way. If you get animals that have not been temporarily kept by the wholesaler, i.e. a direct import, it may also be useful to reduce the germ density of the aquarium with a powerful Ultra Violet device. This can also eliminate many bacteria that may have been introduced and thus prevent the outbreak of epidemics in advance. Professional importers actually have no choice but to use such devices. As far as water quality is concerned, it should be noted that crayfish are very adaptable and tolerate both soft and hard water equally well. Nevertheless, acclimatisation to the new water should be done as gently as possible, so as not to trigger a shock moult in the crayfish. This can also happen if the water changes are too large and too frequent, usually in conjunction with an incorrect high-protein diet. It should also be remembered that it may be necessary, to be able to isolate an inferior or aggressive animal in order to avert losses. Another small aquarium is needed for this. Of course, breeders can also use plastic boxes for this part of the acclimatisation, but these often have the disadvantage, that you can only see them from above. If you want to build up a breeding group, it is not enough to buy just one pair of crayfish. If you want to breed seriously, you have to plan for about 4-6 animals. I would always recommend using 2 males and 2 females or even better 2 males and 4 females. As moulting problems are the most frequent cause of death in crayfish, it can happen at any time that you lose an animal, and it is always very annoying, if it was the "breeding buck" or the "super female", and of course there is no reserve animal available for further breeding. In addition, you cannot tell the exact age of adult crayfish, that you have bought somewhere from the outside, and so it is quite possible that you have acquired animals whose biological clock is just about to run out. For most Australian crayfish, I would plan aquaria with as large a tank depth as possible, with a minimum tank length of one metre. Larger is always better, of course. All crayfish aquaria must be secured against crayfish escaping, as it is normal behaviour for crayfish to migrate to other bodies of water when population densities are too high. Since an aquarium is a very confined habitat, this can happen very quickly. Aquaria with complete covers, top-mounted filters with a suction nozzle in the middle, so that the crayfish cannot get out of it, or masking off the corners of the aquarium where the hoses and cables run with wide strips of adhesive tape ("Mc Gyver tape") have proved successful. Of course, the keeper's daily eye check must not be missing here, true to the motto: "He counted the heads of his loved ones, and behold, instead of six there were seven?" If you have discovered a brood-bearing female, or rather not discovered it, because it has retreated into the furthest corner, then you should consider whether you can carefully transfer the animal to a breeding tank or simply let it go about its brood business, which would certainly be the better way. In my aquaria, it has also proved successful to illuminate them at night with an inexpensive blue light-emitting diode. On the one hand, the crayfish and fish then find their way around the aquarium better at night, and on the other hand, this kind of lighting seems to calm the animals down and thus prevent aggression. Fish can then better avoid approaching crayfish, and losses among the fish are avoided. Unfortunately, there are also fish species, such as the **Banded Rainbowfish** ***Melanotaenia trifasciata***, that cannot be kept together with crayfish and are almost always eaten. This is usually, because these species lie down on the bottom to sleep and then become prey for the nocturnal crayfish at night. And since there is not enough room in the aquarium for them to spread out, the inevitable happens. Therefore, if you wish to keep fish and crayfish together, it is always advisable to ask other aquarium friends for advice beforehand.

Basically, the keeping of Australian crayfish with rainbow fish can be considered ideal, as it largely corresponds to the natural occurrence of these species. Such a biotope aquarium has its own special charm and represents a real challenge.

Artificial roots made of Polyrin cannot negatively influence the water chemistry.

Sex differentiation in Australian crayfish

The males of the Australian crayfish do not have mating gonopods like European or American crayfish, but they do have sex openings at the base of the last pair of their legs and small ischium hooks on them. Some males also have female gonopores, which are said to be non-functional. Unfortunately, it is not yet clear whether they are hermaphrodites. In some species, such as the **Apricot Crayfish *Cherax holthuisi***, the males have somewhat larger scissor legs. In addition, their ***Abdomen*** is usually somewhat narrower than that of the females. The females of the Australian crayfish have two **ovarian openings (*Gonopora*)** at the base of the second pair of their legs, just like the females of the European and American crayfish. In addition, a plate called ***Annulus ventralis*** sits between the last two pairs of legs. The male's sperm packets are attached to this device during mating, so that sometime later the eggs ejected by the female can slide over it for fertilisation. In addition, the females of the Australian crayfish almost always have a slightly wider ***Abdomen* (*Pleon*)** than the males. In some species they also have slightly smaller scissors. Especially with juveniles that are only 3-5 centimetres in size, it can therefore be very difficult to correctly identify the sex. In this case, it would be best to examine an ***Exuvium*** with a magnifying glass.

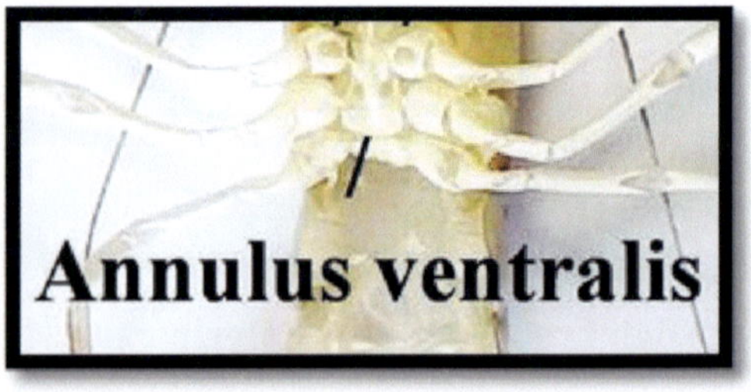

Yabby, *Cherax destructor* Clark, 1936

The Australian **Yabby *Cherax destructor*** is a crayfish that is excellent to breed and keep in the aquarium. This is why this crayfish is also bred commercially in the USA and Israel. A few years ago, there is also said to have been a German company near Detmold that wanted to breed and market Yabbys commercially. Nevertheless, it is very popular among gourmets and is therefore also exported to Europe and the USA. Yabbys can reach a maximum size of about 15centimetres and produce several hundred young two to three times a year. However, they usually only live about 2-3 years in the aquarium. Like the European crayfish, the Australian species are not resistant to crayfish pestilence. Nevertheless, introduced Yabbys could pose a threat to our native species, as they do not require permanently high temperatures like other tropical animals and are very adaptable here. Thus, abandoned animals could well survive the winter in the German or European wild and start to form their own populations. The fact is that Yabbys have already been trafficked to Tasmania, where they now threaten the endemic fauna. In contrast to the fact, that the Yabby is an easily reproducible "farm crayfish", it is also true, that it is now a declining and endangered species in its native Australian habitats. The Australian government has already tried to motivate private individuals to breed Yabbys in their garden ponds. The Yabby can be socialised with all agile or larger fish that do not have veil fins, such as cichlids from Lake Malawi, or rainbow fish like ***Melanotaenia boesemani*** and ***Melanotaenia praecox***. Their name ***Cherax "destructor"*** is probably unjustified due to their distinctly positive social behaviour towards their conspecifics. Only rarely do individual adult Yabbys develop behaviour, in which they seriously injure or even kill weaker conspecifics. Ideally, a breeding group of about 6-10 specimens should be kept. It is best to use aquaria with a tank length of one metre or more. There are several subspecies of the Yabby and some colourful breeding forms, such as i.e. “pearl-blue”.
I assume, that they got the Latin species name **"*destructor*"** (lat.="destroyer") rather, because they can destroy dams and dikes by their digging activities. This is just a natural behaviour that cannot be held against the animals. In the aquarium, they only dig out plants if they are in their way. An aquarium stocked with Yabbys should ideally be set up with lots of root hiding places, stones and a sandy bottom. Floating plants, ***Anubias*** and **Java Moss** are particularly suitable plants. The aquarium may also contain some detritus accumulations, as the broods of the Yabbys need the infusoria living in it. Efficient breeding is not possible in clean and sterile aquaria or rearing tanks. Yabbys are very good at feeding on foliage, dead fish, cooked(!) crab meat and various flake and granulated foods. They often become very trusting and leave their hiding places when food is available. When handling Yabbys, you have to be very careful with the animals, as they

can quickly pinch very sensitively until it bleeds. Large specimens are capable of cracking aquarium thermometers, so these should be placed out of their reach if possible. There are also reports from aquarium enthusiasts that very large Yabbys have even destroyed glass heating elements. Finally, one special feature of the Yabby should be mentioned: Its carapace is very thin shortly before moulting, so that the gastroliths of the Yabby can be seen shimmering through before the moulting. After moulting, the basic colour of the crayfish may change slightly. For example, a blue Yabby of the breeding form "pearl-blue" often looks almost purple. This moulting discolouration disappears after a few days, and the crayfish shines in its old splendour. And in case of a fundamental change in keeping conditions, e.g. water hardness, normally coloured animals may turn to blue.

Blue breeding form of the Yabby ("pearl blue") directly after moulting.

The Yabby *Cherax destructor* in natural colouration.

All morphs of the Yabby have a typical pattern on the upper side of the claws.

Normal colour form of the Yabby, sometimes also traded as "*Cherax destructor* albidus".

Koonac, *Cherax preissii* (Erichson, 1846)

The **Koonac** is a very delicate species that can reach about 20 centimetres in length. The almost black ground colouration and the contrasting orange antennae make this crayfish unmistakable. This crayfish is found in Western Australia, where it also colonises habitats with marine influence. Therefore, it should be tried to maintain these animals with a low salt addition in the aquarium water. The main problem with importing these animals is obviously, that they are transported like food animals at temperatures of 5° Celsius. The real problem is to slowly restart the animals' metabolism after transport. Afterwards, they must not be kept warmer than about 20°. In terms of the general conditions, their keeping would be similar to that of European crayfish. Unfortunately, I have never kept these animals in the aquarium for more than a few weeks; however, I have not yet tried to keep them in a "brackish water aquarium". These animals are therefore not suitable for the average freshwater aquarist, which is why I would strongly advise against buying them. Should it one day be possible to obtain offspring of this species, one should try to find out the husbandry conditions of the breeder so that the species can remain in the aquarium hobby for a few more years because of its very special colourful attractiveness.

Red-Claw, *Cherax quadricarinatus* (von Martens, 1868)

Male Red Claw directly after moulting.

Female after mating with the male's sperm package just attached. The aquarium was set up with plastic plants, as Australian Red Claws usually feed on aquarium plants.

The **Red Claw** reaches a maximum final size of about 30 centimetres, measured with claws. It is also an edible crayfish, that was discovered for aquaristics and reproduces just as productively as the **Yabby *Cherax destructor***. This crayfish species from northern Australia (Northern Territory and parts of Queensland) is also found in the southern part of Papua New Guinea. The Red Claw is popularly propagated as an edible crayfish as it reproduces productively and can actually reach lobster size (weighing up to 600g). The females can produce between 200 and 1500 juveniles per brood and are uniformly blue-green in colour with red spots on the ***Abdomen***. Males also have characteristic red blister spots on their claws from about 6-8 centimetres in size. Despite their size, these animals are peaceful aquarium inhabitants, that can be easily socialised with cichlids from Lake Malawi, for example. Unfortunately, they also eat plants, which is why they should be kept in an unplanted aquarium. Due to their final size, which can be between 20 and 30 centimetres with claws, one has to think about the aquarium size before acquiring them, which should not be less than 1.20 metres tank length if possible. In addition, you should plan in advance where to send the numerous offspring. Keeping them in a perch aquarium is ideal in that the perch always eat a certain proportion of the juveniles, so that you only bring through a portion of the young, which then maintains the stock. These animals should also be fed with beech and alder leaves, which are best watered for a few days beforehand so that the aquarium water does not turn completely brown. When mating, the male throws the female on her back and attaches his sperm packets under her ***Abdomen***. After mating, the female may remain frozen on her back for several minutes. Acclimated aquarium animals also walk around the aquarium during the day. If a female then starts to hide, she should be left in her hiding place, as she is very likely to be carrying eggs. For the young, it is very good to place bricks with many small holes in the aquarium so that there are enough hiding places for the fry. Leaves, detritus and other waste are essential for successful breeding. They grow very quickly and reach sizes of 4-5 centimetres after only a few weeks. You have to offer the young many hiding places, otherwise they will decimate each other. However, they are not as cannibalistic as the American crayfish. It is also advisable to provide the Red Claws specifically with food plants, such as diverse waterweeds and **Hornwort**. **Java Moss** can also be used for the fry. If you catch a Red Claw eating an aquarium fish, you can assume, that it was usually already terminally ill. I have seen it with my own eyes that my Red Claws sat peacefully together in a cave with healthy Malawi cichlids. On the basis of morphological characteristics, such as the male's blister spot or the spines of the tail fan, I would classify the Red Claw as being related to the **"Blue Moon" *Cherax boesemani***. I assume that either one species is directly descended from the other or that both are descended from a common ancestral form. In particular, the attached tissue on the male blister spots seems to be found only in these species. Unfortunately, I have not yet been able to find out whether this tissue has a special function. The Australian Red Claw is a widespread species that shows a very remarkable sexual dimorphism. However, the red blister spots of the males only appear at body lengths of about 6 centimetres or more, sometimes considerably later. This can lead to irritation and wrong purchases. Therefore, when acquiring the animals, one should first examine the underside of the crayfish and look at the position of the gonopores. This is the only way to be sure which sex a crayfish of this species really has. Like the Yabby, the Red Claw has a fine down of bristles on the upper side of its claws, which look slightly furry. Many aquarists thought at first glance, that it was a crayfish infected with the crayfish pestilence ***Aphanomyces astaci***, but this down is only a hairiness typical of the species. It is likely that the animals can smell and taste with these bristles and use them to find food and sexual partners. A crayfish infected with crayfish pestilence, on the other hand, would have white thin threads hanging out of its leg and claw joints. In addition, a crayfish infected in this way would only move very slowly and stagger.

Male Red Claw.

Female Red-Claw.

Lorentz` Crayfish, *Cherax lorentzi* Roux, 1911

This crayfish grows to a maximum length of 12 centimetres and remains significantly smaller than the Red Claw. Characteristic of the Lorentz crayfish are its broader claws, by which it can be easily distinguished from the Red Claw. Its entire body stature is also somewhat stockier. This species needs plant food and therefore also eats aquatic plants. Feeding with foliage and fast-growing aquatic plants such as waterweed is necessary.

Boesemann's Crayfish, *Cherax boesemani* Lukhaup & Pekny, 2008

***Cherax boesemani*, ♂ of the morph "Hoa Creek", traded as "Purple Crayfish".**

***Cherax boesemani*, ♀ of the morph *"Red Brick"*.**

Another striking feature of this species are the long *Antennae*, which clearly exceed the body length. But the most impressive feature of this species is the spiny *Telson*.

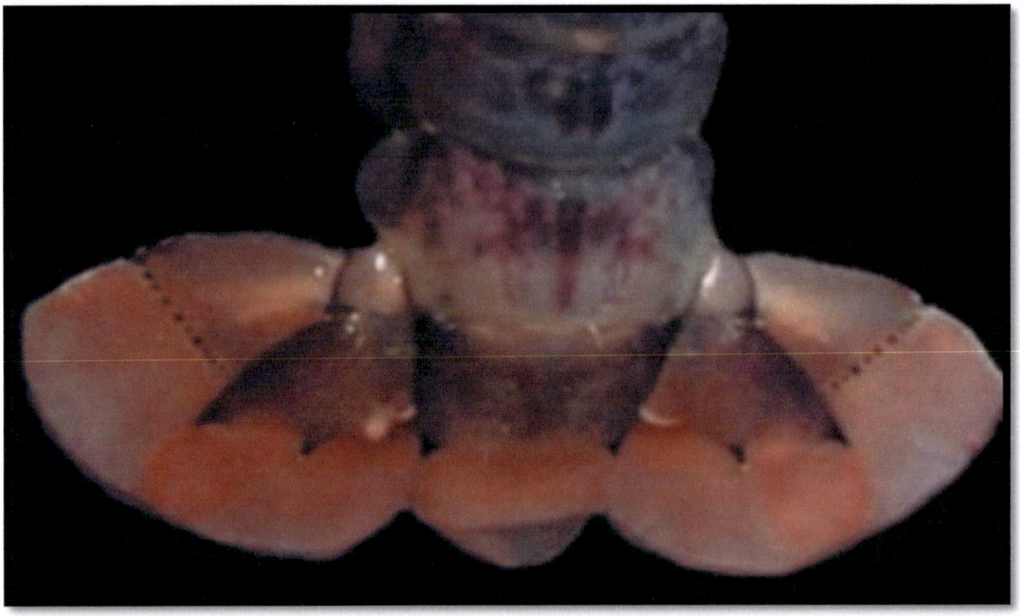

Thorns on the *Telson* of *Cherax boesemani*.

This crayfish already belongs to the group of false Australian crayfish, because it comes from the Vogelkop Peninsula in Papua New Guinea. It grows to at least 10 centimetres long and belongs to the somewhat shy species. The first import shipments contained 80 per cent males, but more females of this beautiful species are now available. Similar to the Red Claw, the males of this species also have blister spots on their claws, but here they are coloured white. Both males and females have white colouration on the tips of the claws, but only males have the attached blisters. Therefore, one has to be very careful here not to confuse sexes. **Boesemann`s Crayfish** is also named and traded as **"Purple Crayfish"**, as ***Cherax spec.*** **"Hoa-Creek"** as well as ***Cherax spec***. **"Red-Brick"**, **"Blue-Brick"** or **"Blue-Moon"**. This means that this species occurs in numerous different morphs. Typical and independent of the colouration are the long *antennae*, which extend beyond the body length, as well as the spines of the ***Telson***. Due to these special characteristics I would therefore also assign the species **"Hoa-Creek",** known as **"Purple Crayfish"** and **"Red-Brick"** to it. I assume that they are merely different coloured morphs of one and the same species. Furthermore, the commercially available species **"Blue-Brick"** and **"Blue-Purple"** can also be assigned to this species. The different colourations could, for example, be due to different food sources or variations in the water chemistry of the different locations, where these "species" are found. In this respect, it should be very interesting to see, how and if the colouration of the offspring still changes. Some breeders have already succeeded in breeding this species and the Purple Crayfish has also been bred. In order to gain final clarity on the species status, these crayfish should ideally still be examined molecularly. It seems likely that this crayfish species was declared as diverse species for purely marketing reasons. **Boesemann's Crayfish** has been officially designated as a new species by Chris Lukhaup & Reinhard Pekny, so that the designation as ***Cherax boesemani*** now reflects the current scientific status.

Alberts` Crayfish, *Cherax albertisii* Nobili, 1899

***Cherax albertisii*, ♂.**

***Cherax albertisii*, ♂ from below.**

This crayfish also comes from Papua New Guinea and can reach a length of 15 and more centimetres. This means it is not as large as the **Red Claw *Cherax quadricarinatus***. What is striking about this species are the very long, slender scissor legs in relation to the body, which make the ***Cephalothorax*** appear somewhat bulky and oversized. This species is occasionally offered in the trade as **Red-Claw *Cherax quadricarinatus***. CHRIS LUKHAUP wrote in 2005, that ***Cherax albertisii*** is genetically identical to ***Cherax quadricarinatus***, so there is considerable doubt as to whether the status of these crayfishes as a separate species is maintained. In any case, because of their striking proportions, these crayfish deviate so seriously from the Red Claw, that they will very probably be classified at least as a subspecies of it in the near future. But we should rather leave this problem to the carcinologists. This crayfish also belongs to the guild of herbivores and should therefore only be kept in aquaria, where plants serve as food, or which have been stocked with plastic plants. The well-known author and publicist Friedrich Bitter pointed out in one of his lectures on American crayfish, that crayfish species, that have long slender claws are more likely to prey on fish than the species with wider claw arms. However, this basic statement cannot be adopted across the board for the Australian crayfish species. ***Cherax albertisii*** could be socialised with angel catfish and king cichlids without any problems, which did not lead to any failures. However, the way crayfish are fed and the size of the aquarium play a decisive role in this question. If the tank is too small, the crayfish may attack the fish, because they have too few possibilities to escape. If crayfish are frequently fed fish meat i.e. smelts, you naturally encourage their taste preferences for fish. That is why I feed my crayfish with plants, with foliage, with high-quality feed granulates and feed containing spirulina. For this reason, it is better to remove dead fish from a crayfish tank, so that you don't give the crayfish a taste for fish in the first place. Breeding ***Cherax albertisii*** is just as productive and problem-free as with ***Cherax quadricarinatus***. The females can carry several hundred eggs, from which the young hatch after about 6 weeks.

Breeding female of *Cherax albertisii.*

Tiger-Crayfish, *Cherax peknyi* Lukhaup & Herbert, 2008

This crayfish was determined to be a new species in April 2008 and is now officially named as ***Cherax peknyi***. It was named by Chris Lukhaup in honour of his long-time friend and global partner in crustacean hunting Reinhard Pekny from Austria. This crayfish is currently traded as **“Tiger”- Crayfish** or **“Zebra”- Crayfish**. This species also originates from Papua New Guinea, from the Fly River system. Previously it was offered as **"*Cherax misolicus*"** or also as **"*Cherax papuanus*"**. This species reaches a maximum length of about 10 centimetres and can optimally be kept as a pair or as a breeding group of one male with several females. Keeping several males in one tank can be fatal for the inferior animals. Especially large males are dominant.

Freshly imported male Tiger Crayfish packed in a plastic box with aquatic plants.

The **Tiger Crayfish** shows a clear sexual dimorphism: While the males have relatively large and bulky claws, those of the females are somewhat smaller. In addition, the ***Abdomen*** of the males is a little narrower than that of the females, but not as extremely different as in European crayfish. However, a reliable sex differentiation can always be made by looking at the gonopores.

Females have much smaller claws and a wider *Pleon*.

Males have larger claws and a slimmer *Pleon*.

In terms of character, the Tiger Crayfish is more of a reclusive and nocturnal species. Breeding is unproblematic, but not overly productive. There are several colour varieties of this species in circulation, which show either more intense contrasts or more orange colour components. This species is not necessarily a herbivorous crayfish, but will dig up plants and roots if they are in the way. Therefore, you have to set up abundantly structured aquaria with a lot of space and many hiding places.

This specimen is very beautifully tinged with purple.

Aquaria for the "Tiger" should have dark substrate and contain many hiding places.

Apricot Crayfish, *Cherax holthuisi* Lukhaup & Pekny, 2006

This *Cherax* species, only recently described and named after the Dutch carcinologist L.B. HOLTHUIS, can reach lengths of up to 12 centimetres and originates from Papua New Guinea. It is very variable in colour and can be grey, beige, blue or orange. Bright orange animals are much sought after in aquaristics, but the import announcements do not always keep their promises. Getting really coloured animals is not always easy, but also not always advisable. For I am of the opinion that when breeding this species, one should use its genetic diversity for sustainable conservation breeding. Therefore, one should also acquire animals that are a little more discreet in colour. The **Apricot Crayfish** is therefore an ideal aquarium crayfish, because it has some very positive attitudes, which make it particularly valuable for aquaristics:

- It does not eat plants
- It does not prey on healthy living fish
- It has striking colours
- It can be kept at room temperature
- It becomes friendly with low fish stocking and good feeding
- It is not aggressive within the species
- It does not need an excessively large aquarium

But first let's start with its herbivorous abstinence. Obviously, this species originates from habitats where there are either virtually no aquatic plants, or it is found in caves. Especially the beige-coloured specimens suggest, that it is a crayfish that is developing into a troglodyte species. The smallness of its eyes also supports this assumption. (The final stage of a troglodyte form would have no eyes at all). Therefore, this crayfish can be fed very well with granulated food, frozen food and foliage. The Apricot Crayfish also exhibits a slight sexual dimorphism, with males having slightly larger claws than females. However, this difference is very slight and it is possible that a male has had to regenerate claws, so the length of the claws is not a reliable sexual characteristic. Unlike other crayfish, the males of this species are very peaceful among themselves. For example, I kept up to ten males in an one metre aquarium, which did not harm each other and even did not attack each other when one animal moulted. Pregnant females retreat and hide in a cave or under root wood. Carrying the brood can take up to ten weeks and requires a lot of patience on the part of the breeder. The Apricot Crayfish is not as productive as other *Cherax*-species and usually only bears one brood per year.

This crayfish can be very well socialised with freshwater shrimps of the *Caridina* and *Neocaridina* genera.

Here is an import with a mixture of blue and orange hues. The blue tones could also have been caused by hard water! This means, that unfortunately one cannot assume that these colours are inherited by the offspring.

For *Cherax holthuisi*, here a male, you can also very well use dark substrate, as this provides a good contrast to the orange colours of this crayfish.

Mating of the Apricot Crayfish can only be observed very rarely, the mating documented here took place during the evening hours. The special thing about this mating was the fact, that the female took the active role and stayed on top of the male. The mating itself lasted just about 5 to 10 minutes.

Juveniles, about one and two centimetres in size.

The juveniles of the orange morph are relatively pale coloured at first, but already get the bright orange colour at a size of two centimetres. Breeding should take place in a tank with a light layer of detritus, foliage and feeding of fine food granules. New imports of this species should be carefully examined to avoid breeding problems later on. This is because they are sometimes infested with crayfish leeches, which sometimes only live as commensals on the crayfish and graze algae there, but sometimes can also cause damage to the crayfish fry.

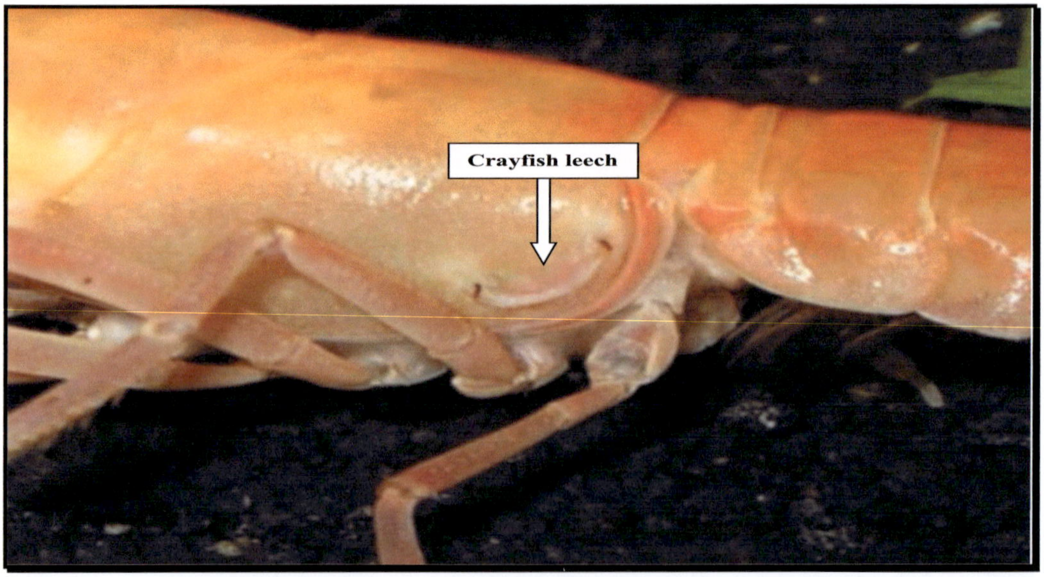

Infected crayfish can be carefully bathed for a few minutes in slightly salty water (e.g. 1 teaspoon of salt to 1 litre of water). With a bit of luck, the leeches will fall off. If not, you can also try very carefully(!) with tweezers.

Male of *Cherax holthuisi* with regenerated claw.

Lost limbs also grow back in the Apricot Crayfish. The lost limb first grows piece by piece as a transparent bag of jelly from the stump of the lost limb. When it is complete, the segments are already visible, but the limb still looks rather rounded and colourless. Only after the next successful moult is the new limb then present again, mostly in its original size. Moulting is a risk for every crayfish and at the same time the number one cause of death for all crayfish.

***Cherax holthuisi* shortly after successful moulting.**

Moulting sequence of an Apricot-Crayfish:

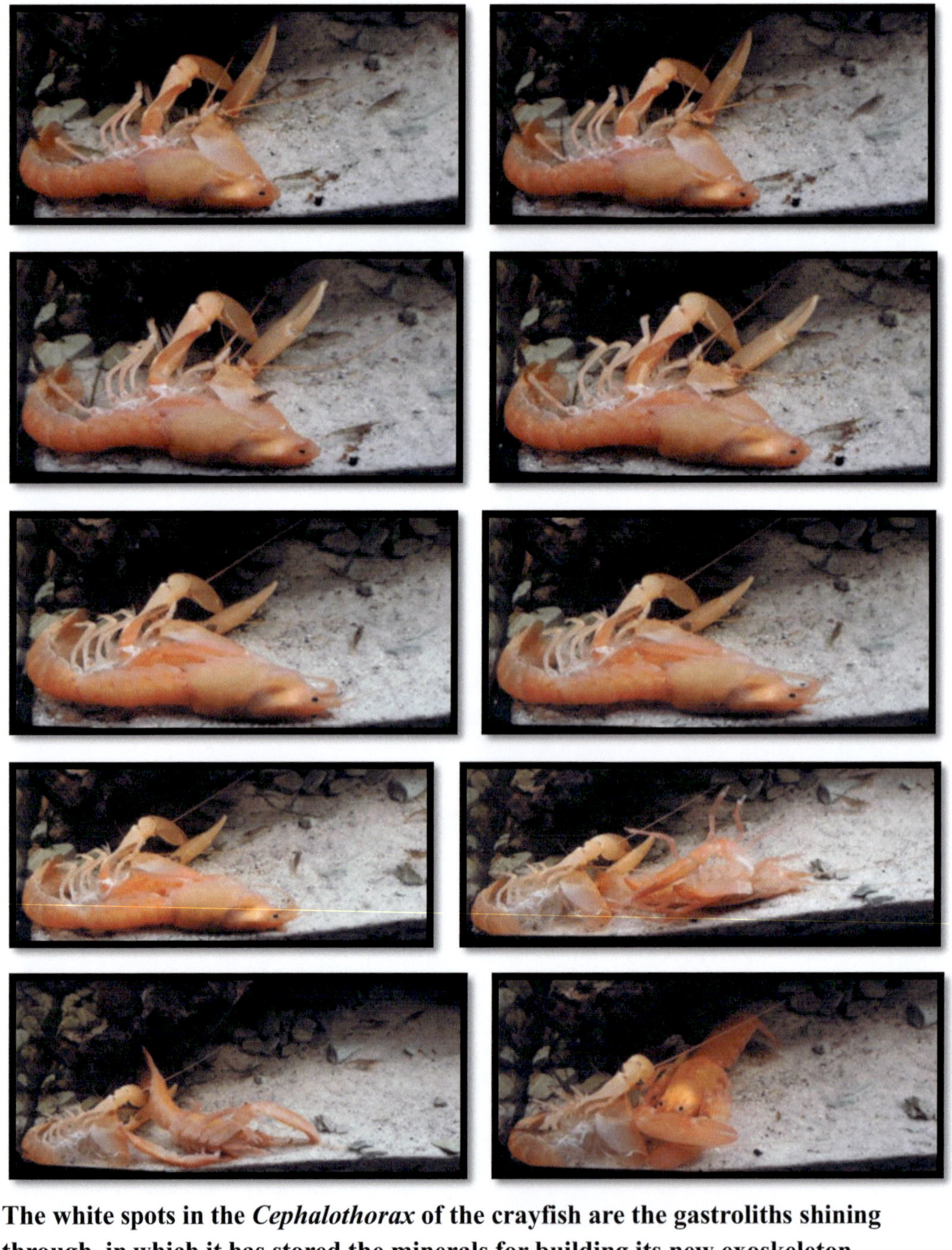

The white spots in the *Cephalothorax* of the crayfish are the gastroliths shining through, in which it has stored the minerals for building its new exoskeleton.

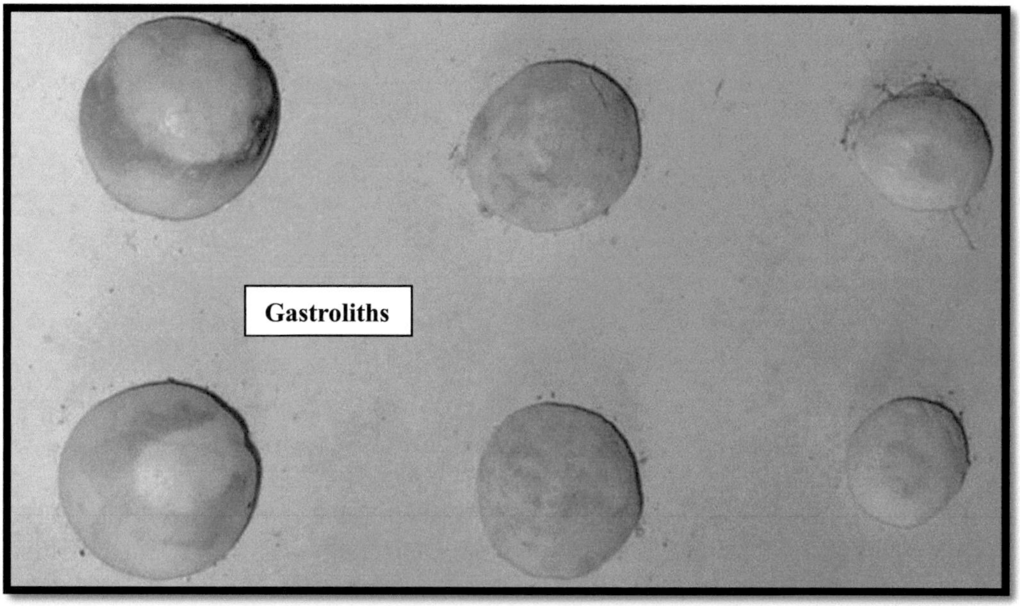

The moulting of crustaceans is a very complicated process, that hasn`t yet been researched in all details. However, it can be said that crustaceans generally control their growth and thus their moulting intervals with the help of hormones (crustecdysone), which they produce in the eyestalks. Experiments have shown, that crayfish that have lost their eyes grow larger throughout their lives and then reach considerably larger final sizes than their conspecifics. Before the actual moult begins, the crayfish extracts the minerals from its shell, which make the shell particularly hard and stores them in the crayfish stones, that it carries around in its stomach. These now grow larger, and shortly before moulting you can see them shimmering through the shell in some crayfish species. Then comes the actual moulting process, where the crayfish opens its exoskeleton at the seam between the ***Cephalothorax*** and ***Abdomen***, and, releasing a jelly-like secretion, begins to slide completely out of the old carapace. In this phase of moulting, the crayfish is completely helpless and defenceless against its enemies, as it cannot flee. Once it has left the shell, it pumps the still soft new shell up a little with water. During this process, an animal of about 7 centimetres length grows by at least 1 centimetre. Afterwards, the crayfish quickly looks for a hiding place where it will stay for a few days until its shell has become firm again. If a crayfish has leeches, they even stay when it moults! Incorrect, by the way, is the assertion that a crayfish must eat its old shell, so that the new one can harden properly. Young crayfish usually moult at intervals of a few days or weeks, adults usually only once or twice a year. The moulting intervals depend on the water temperature and the food supply. During the mating season, entire crayfish populations often moult synchronously.

Family *Astacidae* – European crayfish

Very few endemic species of crayfish occur in Europe, and they also have to deal with many enemies, neozoans and introduced diseases. European crayfish were very abundant in European waters until the 19th century. In the Middle Ages, they were of particular importance as a fasting food, as it was no problem to transport them alive over long distances with the means available at the time. In some places, even churches were decorated with images or carvings of the locally economically important crayfish. My grandmother told me, that her brothers had to go in search of crayfish in the surrounding streams in autumn to contribute to the nutrition of the otherwise rather poor fishing family. Each time they found large quantities, which then had to be cooked and eaten for days until no one could see them anymore. If the poor people living at that time had been told that it was a delicacy, they would certainly have been met with incomprehension.
In contemporary depictions, crayfish were often caught with so-called crayfish plates. A plate-shaped fish trap was fitted with a bait in the middle, lowered to the bottom and reeled in, when crayfish settled on it. But how could it come about, that so common animals, which could be found in droves in every conceivable body of water, whether river, stream or lake, suddenly began to die out population by population? And become an animal, that ultimately had to be placed under nature conservation in order to preserve the few small remaining populations? The culprit, of course, was the "wisest" creature living on this planet, namely *Homo sapiens*, the "wise" human being. But let us first describe the phenomenon of the collective death of crayfish, as it appeared to people at that time: The crayfish, which were actually more active at night, suddenly began to come out of their hiding places during the day and stagger aimlessly over the bottom of the water. Certainly, some contemporaries will have been happy about this and caught one or the other crayfish to use it as edible crayfish. Over the next few days, the movements of the crayfish became slower and slower and more and more apathetic until they laid dead on the ground. In Berlin, there was a wholesaler, who dealt in crayfish and stored them in large ponds for sale. When they suddenly all died of the mysterious disease, an almost indescribable smell must have spread over the whole area. Within a short time, there were dramatic population collapses of crayfish all over Europe, from Italy to Russia. Most large populations were irretrievably lost forever, with only small relict populations remaining here and there. Such as the Weserbergland population, the population in the Harz Mountains or a population in the Pegnitz river near Nuremberg. Today, these are the beacons of hope for the conservation of European crayfish. But how did the sudden catastrophe come about? As mentioned above, this catastrophe is purely human in origin. Several phenomena coincided in this era and caused the collapse of the crayfish stocks. Firstly, during the so-called

founding era, industrialisation was dramatically accelerated. As a result, phenomena such as water bodies and air pollution increased drastically overnight due to the non-existence of sewage treatment or filtration plants. At the same time, rivers and streams were straightened everywhere and other water regulation measures were carried out to optimise the transport of ships and goods. No thought was wasted on any aquatic animals. These interventions resulted in a weakening of the animals' immune systems and a decrease and relocation of various fish stocks. At the same time, new species were introduced, partly intentionally and partly through inland navigation. Towards the end of the 19th century, new crayfish species began to be deliberately introduced from North America. Not only did these spread rapidly, but they also brought with them a deadly cargo for the European crayfish species, namely the filamentous fungus ***Aphanomyces astaci***. This filamentous fungus then triggered the **"crayfish pestilence"**, which almost wiped out the European crayfish populations and led to the phenomena described above. Although American crayfish can also contract crayfish pestilence, they have developed defence mechanisms against the filamentous fungus, that enable them to encapsulate the spores of the pathogen in their body tissue and thus render them harmless. The European crayfish do not have this ability and were thus first infected by the immunised American crayfish and then successfully displaced. Word quickly spread, that the American crayfish were immune to the pestilence, and so more and more American crayfish were released in Europe in the past to restore large healthy crayfish populations. But the new species, such as the **American Crayfish *Orconectes limosus***, usually did not grow as large as the native **Noble Crayfish *Astacus astacus*** and did not have as valuable meat. Until the 1980s, even fishing clubs stocked their angler waters with American crayfish, as these are immune to the crayfish pestilence. Once a body of water has been stocked with these crayfish, it is irretrievably lost for the Noble Crayfish. This was a stocking sin, that is unfortunately often committed. Moreover, some angling clubs actually have the cheek to declare North American crayfish as Noble Crayfish and thus to distinguish themselves as environmentalists at public exhibitions. It is a concern of this little booklet to set the record straight and to promote the preservation of endangered crayfish species for the world of our children through breeding. The biologist Dr. Dietrich Blanke from Lower Saxony, who has unfortunately passed away in the meantime, developed the so-called patchwork strategy for this purpose: Since open water systems connected to water bodies in which American Crayfish already occur, must be considered lost for the Noble Crayfish. Therefore offspring of the Noble Crayfish should be established in as many small compartmentalised water bodies as possible. These primarily include gravel ponds, fishing ponds, garden ponds, ponds in parks and green spaces and similar bodies of water. If an established crayfish stock in such a body of water falls victim to crayfish pestilence, because, for example, water birds have transmitted the spores of the pathogen, a new stocking attempt can be made a few weeks after all crayfish have died. This is, because the pathogen is still active only about 2 weeks after the animals have died. This stocking strategy offers the advantage, that only small crayfish populations disappear in an emergency, but that the species itself is preserved in the large area. In addition, every garden pond owner can make a contribution to nature conservation without having to invest a lot of time or money, provided the size of the pond allows it. My heartfelt request to you as garden pond owners would therefore be, to seriously consider stocking your pond with offspring of the Noble Crayfish. Crayfish are very useful animals that act as health police in a body of water and also recycle leaves and other waste. They and their broods also serve as a valuable source of protein for countless other animals. In a garden pond, they can stop the excessive reproduction of snails. They can also cohabit with newts and other amphibians in sufficiently large ponds. And finally, it should be noted that surplus large specimens can of course also be used as excellent edible crayfish!

This mountain lake is an ideal refuge for species such as Stone- or Noble Crayfish, as it is not directly connected to other bodies of water.

Male Noble Crayfish *Astacus astacus*.

Sex differentiation in European crayfish

The male of the **Long-Clawed Crayfish *Pontastacus leptodactylus*** has significantly larger and longer claws than the female. Due to the loss of one claw and the subsequent regeneration of the claw, it remained smaller, which is why the size of the claws is not a reliable sex differentiator. In contrast, the narrower ***Abdomen*** and the mating pistils, which are in fact rebuilt webbed feet of the ***Abdomen***, are typical.

The female **Long-Clawed Crayfish *Pontastacus leptodactylus*** has a wider ***Abdomen*** **(*Pleon*)** so that the eggs have room there and can be stored like in a pocket. The scissor legs of the females are slightly smaller than those of the males, and at the base of the second pair of walking legs are the ovarian openings through which the eggs are expelled.

Noble Crayfish, *Astacus astacus* (Linnaeus, 1758)

Male Noble Crayfish is searching for food.

Who is crawling? There is no substitute for a daily look at the crayfish!

The **Noble Crayfish** ***Astacus astacus*** once used to be the most common crayfish in Europe. It can reach a maximum length of 18 centimetres in the male sex and live for at least 5-6 years. The females remain somewhat smaller.

The female has smaller pincers and a broad *Pleon*.

The male has larger pincers and a distinctly narrower *Pleon*.

On the underside of the body, male Noble Crayfish have two mating styli at the base of the ***Abdomen***, which are usually carried folded in half between the last pair of legs. Females, on the other hand, have only two small ovarian openings at the base of the second pair of legs. Noble Crayfish mate in autumn, usually in October or November, at water temperatures of about 10-12° Celsius. The male turns the female on her back and attaches a sperm packet between the base of the last pair of legs. A few days to weeks later, the female then expels the eggs, so that they can slide over the sperm package and be fertilised there. The female then stretches her tail fan and sticks the eggs to the webbed feet of her ***Abdomen***. The eggs are fertilised during this procedure and carried around by the female throughout the winter. From May to June, the crayfish larvae hatch out and have to moult twice more under the tail of the female, before they are fully developed little crayfish. In their first year of life, the small crayfish grow fastest and moult up to eight times, if they can grow up in a warm body of water. If, on the other hand, the water is colder, such as a mountain lake, they grow more slowly and moult only five times. Male Noble Crayfish need four years, whereas females need three years to reach sexual maturity. This is one of the reasons why Noble Crayfish are hopelessly inferior to American crayfish species in the same waters, because the American species have considerably faster reproductive cycles. Also, American crayfish species are not as sensitive to water pollution as Noble Crayfish and are often found in silted waters, which Noble Crayfish do not tolerate. Naturally, Noble Crayfish live very secluded during the day and in waters with steep banks they even dig living tubes up to 20 centimetres deep into the banks. Therefore, when keeping crayfish in an aquarium, sand should be offered for digging and roots and stones as hiding places. In sufficiently large aquaria, Noble Crayfish can be socialised with large freshwater fish without any problems. However, Chris Lukhaup and Reinhard Pekny recommended not to associate this crayfish with fish species that sleep on the bottom, such as sun bleaks and common bleaks, as these could fall victim to the crayfish at night. Furthermore, the crayfish should be kept as cold as possible in winter, at temperatures of 10° Celsius and below, so that gonadal maturation can take place. If they are kept too warm during the winter, they may not be able to reproduce. Unfortunately, an aquarium stocked with Noble Crayfish can at best be planted with plastic plants, as they dig up and/or eat real plants. It also makes sense to feed the animals with watered beech or oak leaves. The animals also accept earthworms, dead fish, boiled(!) crab meat and granulated food. Of course, the food should always be given in moderation, so that no foci of putrefaction form in the corners of the aquarium. Vigorous water circulation with strong currents is also useful, especially in summer when the oxygen content of the aquarium water drops as temperatures rise. Crayfish, that try to climb to the water surface, are usually indicating, that something is wrong with the oxygen content or with the general water chemistry (such as nitrite or other toxins in the water, etc.). Crayfish are generally very adaptable animals that can tolerate both hard and soft water and both high and low pH values well if they are slowly acclimatised to new water values. Large male Noble Crayfish often dominate other males and can also be dangerous to them. As they cannot easily avoid each other in the confines of an aquarium, cannibalism can lead to losses. This can be counteracted by feeding the animals a varied and balanced diet and by observing the animals and, if necessary, moving endangered animals that have just shed their skin elsewhere in good time. However, it is best to keep only one male in an aquarium, which you can socialise with two or three females. Boiled crab or shrimp meat can be fed to crayfish very well, as it contains exactly the proteins these crustaceans need. Another advantage of boiled meat is that the proteins have been bound by the boiling process in such a way, that they do not overly pollute the aquarium water and can thus even lead to poisoning of the water by ammonium and ammonia. In addition, boiling kills harmful germs that are usually lethal for the crayfish. In particular, the spores of the already mentioned crayfish pestilence should be mentioned here, but also the teliospores of the so-called porcelain disease, with which crayfish can become infected by eating dead conspecifics. In order to exclude the possibility of infection with this disease, dead crayfish should always be removed immediately from the aquarium or garden pond. Animals with porcelain disease can be recognised by the fact, that they do not have transparent, but white body tissue under the ***Abdomen***. These crayfish move a little slower and can be caught easily. Another disease is rust spot disease, which usually affects weakened animals. This is caused by a virus and brings out rusty brown spots on the shell of the crayfish. These can also heal, but the disease can lead to the death of the crayfish, if the spot is in a part of the carapace that the crayfish cannot regenerate properly during its moult. In the worst case, the crayfish cannot get out of the old carapace, because it is virtually stuck at the spot with the stain. Newly acquired crayfish should definitely be kept in a quarantine tank for a few weeks to minimise the risk of disease for the existing stock. If you know, how dramatic a crayfish pestilence epidemic can wipe out an entire population, you will be very wary of putting crayfish together with new acquisitions without hesitation. Therefore, you should always take enough time for the constant observation of your animals, even if your partner might get jealous. In case of doubt, carelessness can cost you the entire stock!

Eurasian Minnow, Sterlet Sturgeon and Golden Orff are fish species that can be socialised with large Noble Crayfish in a suitably spacious cold-water aquarium.

This male Noble Crayfish takes cover under a scour.

Long-Clawed Crayfish, *Pontastacus leptodactylus* (Eschscholtz, 1823)

Male Long-Clawed Crayfish.

Female Long-Clawed Crayfish with eggs under its *Pleon*.

A Long-Clawed Crayfish in typical defensive posture with open scissor fingers.

The Long-Clawed Crayfish can sometimes also appear as a blue morph. However, this colouration is only passed on to the offspring if the antennae are also blue in colour. Blue specimens with normally coloured antennae only produce normally coloured broods. In these specimens, blue colouration usually occurs as a result of water chemical conditions. For example, when the water is particularly hard and calcareous.

The **Long-Clawed Crayfish** is not just one species, but a complex of diverse forms and subspecies, whose status is always highly disputed. Therefore, only a very simplified presentation of the current taxonomy is given here. The Long-Clawed Crayfish also occurs in brackish or swampy habitats which are too muddy for Noble Crayfish. When introduced into Noble Crayfish waters, it displaces them. It also reaches enormous sizes, whereby the males in particular can develop extremely large claws. Total lengths of males up to 18 centimetres are possible, but their total weight is less than that of Noble Crayfish of the same length, because of their slimmer structure. This relative of the Noble Crayfish is found mainly in south-eastern Europe and Asia. Males have considerably larger claws than females and a much narrower ***Abdomen***. Otherwise, the sexes are distinguished in the same way as in the Noble Crayfish. They are mostly imported from Iran or Turkey as edible crayfish, but the quality of their meat is lower than that of Noble Crayfish. There are now also naturalised populations in Germany. They can be easily distinguished from Noble Crayfish, because their ***Cephalothorax*** is covered with rough spines, whereas Noble Crayfish have a smooth one. These crayfish mate at water temperatures of around 12° Celsius, but their mating season is not necessarily in autumn as with the Noble Crayfish. I myself was able to observe a mating in spring; the eggs were released two days later. This mating occurred when I took a female and a male out of a pond and put them together in a bucket. A quarter of an hour later, the male had thrown the female on her back and fertilised her. This mating was purely accidental. So I cannot say, if the experiment can be repeated at any time. The development of the eggs takes about one month at water temperatures of 12-15°. If the temperature rises too quickly, the fry dies. Egg-bearing females from imports therefore usually produce only very few juveniles. In aquaria the animals are peaceful among themselves, if the stocking density is not too high. They can be fed well with smelts, food granules and foliage. Socialisation with most fish is unproblematic. They also tolerate higher temperatures, but need a cold season for breeding. According to literature they can live for about 6 years. Contrary to claims to the contrary, the Long-Clawed Crayfish is not immune to crayfish pestilence, which I can confirm from my own experience. From the grocery store I got some Galician crabs during my school days, which looked healthy at first. Then after a few days you could see thin white threads on the claw and leg joints that moved in the current. Then the animals became more and more apathetic and finally all died. This is why you must always quarantine newly acquired crayfish. Animals from the food trade should first be given a cold shower. Then slowly increase the water level so that they gradually get used to breathing in water again and can "breathe out" air that has accumulated under their shell. After 2 days it is usually decided which animals have survived the hardships of the food trade and which have not. Ideally, you should take the smallest possible crayfish charges for aquaristics.

Signal Crayfish, *Pacifastacus leniusculus* (Dana, 1852)

One-size-fits-all juvenile, about 6 centimetres long.

***Exuvium* of a large Signal Crayfish male, 20 centimetres long. This animal came from the hatchery of Rüdiger Everding. Unfortunately, the legs shifted a little backwards during drying, which is why it is better to fix exuviae when drying.**

The **Signal Crayfish** also belongs to the *Astacidae* family and thus to the closer relatives of our native Noble Crayfish *Astacus astacus*. It can reach sizes of 18 centimetres and a weight of up to 300 grams. These animals look confusingly similar to the Noble Crayfish, but have distinctive white spots on the joints of the clawed fingers. The claws are bright orange when viewed from below, which is where the animals got their German name. The animals originally came from the Sacramento River system in California and have acclimatised very well in our country. It is still controversial whether these animals themselves are sensitive to the crayfish pestilence or whether they transmit it. The fact is, however, that in Sweden and Austria, where they were released into waters with Noble Crayfish populations, they have mostly displaced the Noble Crayfish. During a summer holiday at a lake in central Sweden, I found several exuviae and dead specimens of the signal crayfish, but compared to the 30 or so sightings of signal crayfish tracks, I only found one dead Noble Crayfish. From this I conclude that Noble Crayfish and signal crayfish can certainly exist together in one body of water, but that the signal crayfish will gain the upper hand in the long run. Fortunately, they have been relatively little distributed in Germany; however, they have been and still are bred as edible crayfish. The same applies to them as to the crayfish and other imported crayfish species: they have no right to exist in our waters. In the aquarium, signal crayfish keep well, but they need the largest possible aquaria with lots of hiding places and should be kept in pairs or harems. The aquarium must also be designed to be as escape-proof as possible, as they are masters at breaking out and are excellent climbers. Plants serve as food and should be offered to them as such. Hornwort, spring moss and water plant are best suited for this purpose. From the signal crayfish breeder Rüdiger Everding I learned by personal communication that in the past cross-breeding experiments of signal crayfish and Noble Crayfish were carried out. In fact, a copula with subsequent egg-laying did occur, whereby the eggs initially looked fertilised but died after a few days or weeks. It is a comforting thought to know that the invaders from North American waters are not joined by hybrid crayfish that can later hardly be clearly assigned to any species. Once released, such genetic mixtures are almost impossible to clean up and leave behind a scorched earth for the gene pool of the affected unadulterated pure species that cannot be restored. The claw joints of the signal crayfish shine conspicuously white (see picture below right); in some publications signal crayfish have unfortunately been wrongly depicted as Noble Crayfish(!)... However, crayfish never have these characteristic white spots on their claw joints!

The crayfish pestilence - chronology of an epidemic

First appearance in Europe:
Around **1860** Lombardy in northern Italy. The epidemic then continued as follows:

1875 France, **1880** Central Europe,**1890** Russia, **1893** Finland, **1900** Bulgaria, **1907** Sweden, **1978** Spain, **1981** British Isles, **1982** Greece, **1985** Turkey and **1987** Norway.

During these major events, entire populations of endemic European crayfish disappeared within a few months; Noble Crayfish populations were not always affected. In Turkey, for example, populations of the Long-Clawed Crayfish were also wiped out. In retrospect, we can now say that about 95%(!) of European crayfish stocks were destroyed. Most of them without any chance of recovering! In addition to these major events, stocks in small waters were also hit by the crayfish pestilence. Only some few residual populations of European crayfish have survived in largely isolated areas, which don`t have any connections to water systems with introduced American crayfish species. But what is "crayfish pestilence" and what are the effects of the pathogen? Crayfish pestilence is the tubular filamentous fungus *Aphanomyces astaci* from the oomycete family (egg fungi), which uses American crayfish species as hosts and reproduces via these hosts. In a healthy American crayfish, infestation with the zoospores of the fungus does not lead to the outbreak of crayfish pestilence, because the crayfish encapsulates the zoospores of the pathogen in its cuticle (its outer skin) with the body's own enzymes, thus preventing the hyphae (filaments) of the fungus from penetrating its body and damaging vital organs. In the process, the infected crustaceans form melanin, i.e. black spots, in the affected areas. The zoospores of the fungus mainly attack soft parts of the crayfish's body, such as tissue on the eyestalks, the conjunctiva between joints and similar, as they can penetrate the crayfish's body more easily here. For the American crayfish, this is not too tragic as long as its immune system is intact. However, if the crayfish's immune system is attacked by factors such as polluted water or an extremely high population density, it too can contract crayfish pestilence and die. European crayfish have not yet developed this immune defence, and therefore have no chance of survival, if they come into contact with the zoospores of the pathogen. However, there are faint signs of hope that European crayfish have begun to build up resistance to the insidious tubular filamentous fungus. The first is, that European crayfish also emit melanin when they are infected with the pathogen. However, the amount of melanin is not yet sufficient to prevent the pathogen from attacking vital organs. Since the mortality rate of European crayfish is currently still around 100%, it can hardly be assumed, that crayfish resistant to the pathogen will survive in the short term. However, there are now a few sympatric occurrences of European and American crayfish, which indicate that the crayfish living in these water systems will build up resistance to *Aphanomyces astaci* in the medium to long term. For example, there are common occurrences of Signal Crayfish and Noble Crayfish in central Swedish lakes. And a common occurrence of North American Crayfish and Noble Crayfish at a German dam in the Harz Mountains. At the present time, however, no final statement can be made as to whether these joint populations can coexist permanently or whether the great total loss of European crayfish will occur at some point. Two things are problematic here: The first is, that American crayfish, even if they themselves are not diseased, can transmit the disease at any time simply by moulting, since the encapsulated fungal spores are released during a moult. The second thing is, that the spores of crayfish pestilence can be introduced not only by crayfish, but also by fish, objects and even water. The fungal spores themselves can also attach themselves at least up to four times to false hosts, until they have finally found a suitable host from which to multiply further. In this process, they also benefit from the crayfish's ability to act as the health police of the water body and to utilise dead conspecifics as food. If crayfish have been infected with crayfish pestilence, one day after the infection it can be observed, that the crayfish try to clean themselves with their legs all over their body; obviously, the zoospores cause a strong itching at first. This is followed by a phase in which the animals leave their hiding places even during the day and run around staggering and uncoordinated. Then they start shedding limbs by autotomy. Finally, they become increasingly limp and die. These processes usually take place within a few days, depending on the water temperature and vitality of the infested crayfish. So far, there is no treatment against this disease, except for preventive hygienic measures (see also chapter on pond hygienics). If the zoospores of the pathogen do not find any new hosts, they die after about 15 days. In concrete terms, this means that a crayfish stock destroyed by crayfish pestilence can be restocked after a certain quarantine period. However, one should bear in mind, that crayfish pestilence in sparsely populated crayfish waters can also progress over longer periods of time and rather insidiously, if the zoospores rarely encounter a host. Dead crayfish should therefore always be removed from the water and, if possible, boiled or burned to successfully contain the disease. Remaining crayfish that survive the pestilence should be critically observed and examined by veterinarians to stop further transmission of the disease. Since the zoospores can also attach themselves to fish, the disease can also spread upstream. If a body of water is infected with crayfish pestilence, the entire body of water should therefore - if possible - be sealed off from other bodies of water. Under no circumstances should crayfish from an infected body of water be moved to other bodies of water!

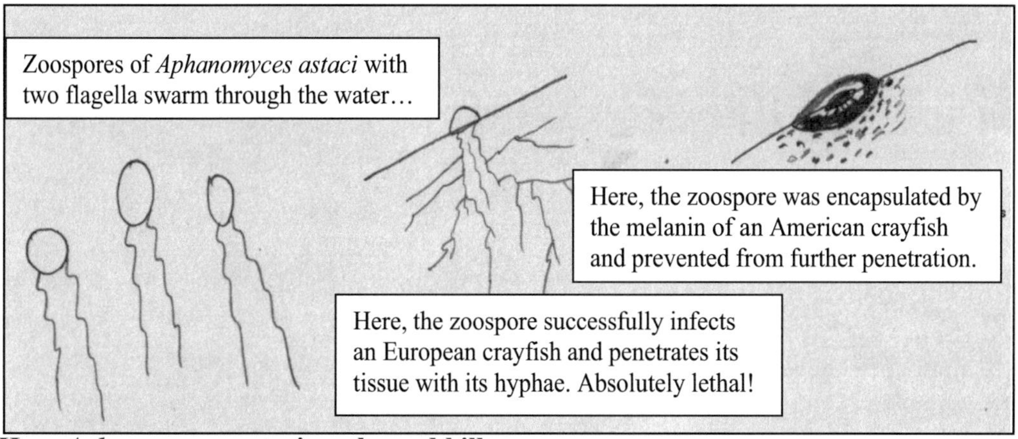

How *Aphanomyces astaci* works and kills…

When I was at school in the 1980s, we bought Long-Clawed Swamp Crayfish from the delicatessen trade for our biology study group. However, they died of the crayfish pestilence after only a few days. We couldn't save a single one of these animals!

Family *Cambaridae* - Crayfish from North and Central America

The representatives of this family are found in North and Central America as well as in East Asia. They are one of the most species-rich crayfish families, of which there are very different representatives. They have conquered a wide variety of habitats and some species have developed such specialised lifestyles, that they can only be found in very specific places. Some are threatened with extinction in their native biotopes, because within North America, various crayfish species have been moved to other aquatic systems by human impact, or because their biotopes have been destroyed by human influence. Only very few species have been popularised for aquaristics in relation to the species present. Of these species, most would probably have to be classified as ecological pests, as they endanger other endemic crayfish all over the world. The natural forms of these species usually do not look very colourful and show rather brown-green shades, but colourful animals have been bred especially for aquaristics, with the range of shades extending from orange to blue to brilliant white. Therefore, one should always differentiate between natural forms and cultivated forms and be aware, that cultivated forms only show natural behaviour to a limited extent. It is thanks to Friedrich Bitter and Lutz Döring, that a few non-commercially colourful species of this family were brought to Germany from North America, where they are being bred by dedicated aquarium enthusiasts. However, breeding them is not without problems, as they need a cold period in winter for the maturation of their gonads, just like the European crayfish. Also, these species are not as productive as the species presented in this book, which are bred in crayfish farms and some of which are distributed worldwide. It would be an interesting challenge to recreate a North American crayfish biotope with the appropriate fish species. Fish species such as Pumpkinseed, North American Orff and some other species are suitable for this, but unfortunately are sometimes not available. It is also possible, to keep viviparous tooth carps here, as some of these species originate from the Mexican highlands and can therefore be kept and bred well at room temperature. These include in particular Guppies, Platies, Swordtails and various species of Mollies and Sailfin Mollies. Some of these species even tolerate brackish and salt water. Unsuitable, however, are all breeding forms with long veil fins, as these can be preyed upon by the crayfish. Also unsuitable are the representatives of the North American catfish, as the latter kill and eat the crayfish at the latest during moulting. Finally, it should be noted, that the trade and reproduction of some American crayfish, such as the **Red Swamp Crayfish *Procambarus clarkii***, i.e., has now been banned in the European Union, because these species endanger endemic autochthonous crayfish species in the wild by displacing them and transmitting crayfish pestilence.

Red Swamp Crayfish, *Procambarus clarkii* (Girard, 1852)

Here a male in an aggressive defensive posture with raised claws.

Male of the Red Swamp Crayfish cleaning their eyes, here the orange morph.

The **Red Swamp Crayfish** originates from Florida and has since spread worldwide. It reaches a total length of about 12 centimetres and is bred as a farm crayfish in various countries on different continents. A by-product of this farm crayfish are very colourful morphs, which are sometimes traded at very high prices in the aquarium trade. In the meantime, there are even populations on the main African continent, where there are actually no crayfish naturally, and in the course of globalisation there is hardly a country on earth, that is safe from these adaptable animals. They also migrate long distances over land, they damage plantations and they undermine dykes and dams. The damage they cause is enormous, and this aspect of their occurrence alone should not be underestimated. Like other North American crayfish species, they also transmit crayfish pestilence and make reintroduction of native crayfish species impossible in waters they have once opened up. Since these animals often appear in the pet trade, here is an urgent request to all aquarium enthusiasts: Please do not keep or release these animals in the wild under any circumstances! A single escaped pregnant female could cause a similar catastrophe as the 100 North American Crayfish, that were released in the Mietze river near Berlin at the end of the 19th century. *Procambarus clarkii* is a very productive species that can have 50-300 offspring with only one brood, depending on size. When kept warm, they reproduce up to three times a year. Within the species, this crayfish is territorial and therefore quite aggressive in the limited space of an aquarium. The juveniles are also already very cannibalistic and need a lot of space and food. It is best to keep pairs or breeding groups with one male and several females. The sex differentiation of the species of this crayfish family corresponds to that of the European species. White, orange and blue morphs of this crayfish are now available for aquarists. Such crayfish kept in aquaria often develop their own personality and are not nearly as shy as some Australian crayfish species. They often even threaten with their claws if a curious person comes too close to the aquarium. They are animals, whose keeping can really excite the crayfish keeper. Because of their interesting biology, they are also ideal study objects for schools and school projects. With this species, children can already be given a comprehensive knowledge of crayfish and the associated environmental issues. Responsible aquarium enthusiasts should always make sure, that these animals do not get into the wild under any circumstances, as some populations of this species have already formed in Africa and Europe, endangering native species. It is imperative to prevent them from spreading even further in the wild - with the well-known ecological consequences.

Blue morph of *Procambarus clarkii* during copula. This can take a relatively long time, sometimes even up to an hour.

Since the **Red Swamp Crayfish** is extremely adaptable and survives our winters without any problems, this species poses a very serious threat to our endemic fauna. In addition, it has also been proven to destroy rice fields and dykes, and in the areas where it was introduced, it also becomes a nuisance to the people living there. It would not be surprising, if the keeping and breeding of this species were to be regulated or banned by many states in the foreseeable future.

North American Crayfish, *Orconectes limosus* (Rafinesque, 1817)

The **North American Crayfish** can be recognised by the characteristic red transverse bands on the segments of its ***Abdomen*** and the orange-coloured tips of its claws. (Unfortunately, the colouration of the tips of the claws has not been preserved in the preserved specimen shown here). It reaches a maximum length of about 12 centimetres and has considerably less usable meat compared to the Noble Crayfish. This crayfish was released in northern Germany more than 100 years ago in the 1880tes. Only about 100 animals were released into the small Mietze River near Berlin. Today, it is assumed that almost all northern German crayfish of this species can be traced back to this one import. The population has certainly grown to millions of animals. They colonised all kinds of waters and make the reintroduction of Noble Crayfish impossible, because they act as carriers of the crayfish pestilence. On top of that, they were released into various fishing ponds by angling clubs in the seventies and eighties of the 20th century, so that no more Noble Crayfish can be stocked on these ecologically valuable areas. On top of that, the angling clubs have issued protective measures (!) for this species, so that this changing ecological disaster can spread even better and is not "overfished"! The fact, that even fisheries authorities cooperate in such an environmental outrage, is actually a scandal of immense proportions. Just imagine that the Ministry of Agriculture would set up protection zones for Colorado Potato Beetles, so that they could find non-poisoned breeding grounds - then you have roughly understood the absurdity at stake here. If you wanted to act as a true conservationist, you would actually have to collect these animals, and - no matter how big or if carrying brood - put them in the boiling pot! Under no circumstances should these animals be spread or protected in any way, as they have successfully infiltrated almost all kinds of water bodies in Germany, including channels, rivers and many lakes. North American Crayfish can be kept very well in cold water aquaria or in unheated indoor aquaria. As they have relatively low life requirements compared to other crayfish species, they are easy to keep. They eat all common types of food and should not be placed in planted aquaria, as the planting is only food for them. North American Crayfish can be found from April to September in the shore zone of gravel ponds or between stones on canal banks, hiding during the day under the cover of plants, wood, roots or stones. In some bathing lakes in summer, they even fall victim to bathers, who step on the animal's half buried in the mud and probably do not even notice it. I have already seen one victim of such an accident with my own eyes (the crayfish, not the bather!). But the crowning glory of this environmental scandal is put on by fishing clubs, that put this crayfish in their club waters, because it is immune to the crayfish pestilence (which is only transmitted by the crayfish itself.). A cynicism, that cannot be surpassed. At this point, one could also imagine a business enterprise only hiring employees, who are HIV-positive, with the justification, that they could no longer be infected with AIDS... What madness! If the spreaders of crayfish were held liable for their stocking actions, they would certainly have to pay several million dollars in compensation to make up for some of the ecological damage. This little booklet should not contain any political aspects, but in view of the explosive nature of this issue (also in monetary terms), I simply cannot refrain from mentioning these small side aspects. I have strong hopes, that a growing environmental awareness in our population will advance the cause of crayfish in the long run. In summary, it can be said, that the introduction of North American crayfish species into foreign waters have almost completely ruined the populations of native crayfish species. Also, the North American Crayfish is neither a valuable edible crayfish, nor an indicator of good water conditions, as it can survive well in muddy and polluted waters. Moreover, it is certainly a carrier of other parasites, which is why it is better to beware of eating them. Such parasites get into the crayfish mainly by eating water snails, through which they then regularly get into waterfowl under sub-normal circumstances. There, these parasites finish their life cycle by infesting the internal organs of the birds and producing their eggs, which are then excreted by the birds back into the water. Fortunately, infection with such parasitic organisms is rare in humans, because most people boil their food before eating it. If, on the other hand, one eats raw crayfish in the manner of sushi, one accepts infection with life-threatening parasites, which, in the absence of a suitable host, can now cause very serious damage to any vital organs. Infirmity and funny doctor's visits are then guaranteed!

Blue Florida Crayfish, *Procambarus alleni* (Faxon, 1884)

***Procambarus alleni*, here a female.**

The **Blue Florida Crayfish *Procambarus alleni*** originates from Florida. It usually reaches sizes of 8-10 centimetres body length and should be kept in pairs or harems. This crayfish is not quite as aggressive as the related **Red Swamp Crayfish *Procambarus clarkii***, but should still be kept under close observation to prevent losses. Breeding is unproblematic as with other subtropical American crayfish. Usually a blue form is sold in the aquarium trade, as the brown natural form is not attractive enough. The animals like to eat detritus with the small animals living in it and should therefore be kept in tanks with foliage and food plants. They can be easily distinguished from *Procambarus clarkii* by the fact that they have considerably smaller warts on the upper side of their claws. Also the shape of the claws of adult males is proportionally somewhat narrower than those of *Procambarus clarkii*. They can be socialised in the aquarium with fish, which do not sleep on the bottom, and should ideally be kept in a tank including some detritus. The crayfish fry also need the infusoria contained in the detritus and therefore cannot develop in sterile rearing tanks. The Blue Florida Crayfish likes to dig up plants, which is why floating food plants such as waterweed and hornwort are more suitable here.

Marbled or Self-Cloning Crayfish, *Procambarus cf. fallax* (Hagen, 1870)

The **Marbled** or **Self-Cloning Crayfish** grows to a maximum size of about 12 centimetres, but usually remains smaller. It clearly belongs to the ***Cambaridae*** and the genus ***Procambarus***, but the species could not be officially identified yet, as there are only females of this crayfish species, which were mainly distributed by aquaristics. Since taxonomic rules require a male for an exact species determination, taxonomists cannot do much with this "female" species. Research has shown, that this crayfish reproduces by parthenogenesis, i.e. by virgin reproduction. This means, that the species is able to clone itself into its brood by a genetic trick, that we do not yet know exactly. Normally, every crayfish has a single haploid set of chromosomes, which usually must be fused with another haploid set of chromosomes from a sexual partner to form a double set, i.e. a diploid set of chromosomes. The Marbled Crayfish has invented a trick to circumvent this law and duplicates the chromosome sets in its eggs itself. Therefore, they do not need to be fertilised. Marbled Crayfish are definitely not hermaphrodites or sex-changers. But two things are a fact: Marbled Crayfish never mate with conspecifics, and they stop reproducing automatically as soon as the population has grown too large. Thus, not only can they reproduce without a sexual partner, but they can even exert direct "birth control".

Individually placed marble crayfish usually start carrying eggs shortly after transfer. This is a very sensible and resource-saving reproductive strategy. Based on my many years of keeping experience and observations of this crayfish, I would like to formulate the following hypothesis regarding its species status: Based on the species descriptions and photos provided by Chris Lukhaup and some other circumstantial evidence, I would want to assign this **Marbled Crayfish** close to the species ***Procambarus fallax***. Justification: The *Procambarus fallax* pictured and described by Chris Lukhaup looks most similar to the Marbled Crayfish in terms of phenotype; especially in body grain and striation of the ***Pleon***. According to Chris Lukhaup, *Procambarus fallax* is originated in Georgia and Florida, which may have strongly favoured the possibility of exporting these crayfish to other parts of the world. It is also possible that the Marbled Crayfish were only exported by chance as by-catch of *Procambarus clarkii*, which also originates from Florida. The aquarium professional Karl Zöllner told me, that Marble Crayfish are said to have reached Germany via Shanghai as early as the 1980s. This circumstance would also support the thesis, that these animals came to Asia as by-catch from American farmed crayfish. The Asian breeders certainly quickly realised, how easy it was to reproduce this species, and so it was first bred for food production and only later for aquaristics. Personally, I would not be surprised, if the species was exported from the USA to Asia much earlier, possibly as early as the 1960s or 1970s. But why are there only females? Shouldn't males have been exported as well? There are several theories on this. The best I could imagine is, that under normal environmental conditions the Marbled Crayfish occurs in both sexes in nature. However, if its habitats dry out for certain periods of time, the sex partners do not have the opportunity to produce offspring together due to their environmental conditions. Either the risk of drying out on the way to the sexual partner is too great, or the general enemy pressure is so high during drought, that the crayfish cannot leave their holes. Chris Lukhaup wrote about *Procambarus fallax,* that he was able to excavate some specimens of this species from holes and tunnel systems. That would support my theory at this point. Now, if a short-lived species (in nature there are no conditions without enemy pressure) can improve its survival risk in such an environment by a genetic trick, this means of course a great advantage over other species without such tricks. It is also conceivable that the males of *Procambarus fallax* are quite short-lived fellows, who cannot survive dry periods of months as well as the females. Thus, the females must somehow compensate for the lack of males. For sure, some will still carry broods that were also fertilised by males, but others will then only carry self-fertilised eggs in extreme dry periods. I think, that the early imported Marbled Crayfish were just such animals, compensating for the loss of males at the time of collection. Possibly, for reasons we do not know today, the crayfish catchers only collected egg-bearing females. Since the collected animals were only true self-fertilising females, at this point their regular normal line of reproduction was interrupted and thus only their "emergency programme" continued to be bred by humans. I have thought about the problem of the Marbled Crayfish for many years, but have now come to the conclusion that the hypothesis I have described seems very valid. For example, Chris Lukhaup states small streams and rivers, but also stagnant waters and small ponds as habitats for *Procambarus fallax*. This statement about the occurrence of this species would thus exactly support my hypothesis of the adaptation of the Marbled Crayfish to temporary waters through parthenogenesis. The worrying thing about this fascinating crayfish is that it too can be a carrier of crayfish pestilence. Moreover, even a single animal that escapes into the wild, can establish an entire population! I would not be surprised, if this crayfish one day succeeds in displacing the introduced and now common **North American Crayfish *Orconectes limosus***. So far, this species has already been detected in some small German waters; a spread and dispersal by water birds, but also by ignorant humans, can no longer be ruled out. It is a fact, that as North American species they survive many climates, and that even cold winters do not bother them. In aquaria, these crayfish can be kept and reproduced very well, and I would even advise keeping them as food animals or as study animals for schools and universities. However, they eat all the plants and must also be fed on a plant diet. They are quite peaceful towards most fish and, as mentioned earlier, they will stop reproducing if their population becomes too large. With this ability, under natural conditions, the animals will certainly curb somewhat the general juvenile cannibalism that many crayfish unfortunately display. Another advantage that serves the survival of the species. A final note: Because of its mode of reproduction and its numerous problematic characteristics, this species may no longer be offered at various aquarium fairs.

Cuban Crayfish, *Procambarus cubensis* (Erichson, 1846)

The **Cuban Crayfish** actually comes from Cuba and some neighbouring small islands. It reaches a maximum length of about 7 centimetres and is thus also suitable for keeping in smaller aquaria with a tank length of 60 to 80 centimetres. The species also tolerates slightly alkaline water with a somewhat higher total hardness, which suggests, that the Cuban Crayfish can also be found in natural habitats with a certain amount of brackish water. The long slender scissor fingers of this species suggest, that it could be a skilful fish predator that preys on fish resting on the bottom, at least at night. This crayfish also eats plants, so it is better to use floating plants and hornwort for tanks with this crayfish. Otherwise, this species is a frugal omnivore, which can be fed in exactly the same way as other species of this crayfish family. Kept warm, the species is also productive and can produce offspring several times a year. Recently, this species has become a little quieter, but it was first successfully imported and bred towards the end of the 1990s. However, as it could not keep up with most other crayfish in terms of colour, it has probably gone a little out of fashion. Due to the origin of these crayfish, I would recommend socialising them with live-bearing tooth carps such as guppies, platies, mollies and swordtails. Whereby on the one hand the crayfish then occasionally prey on old and weak fish, while on the other hand the fish decimate and regulate the crayfish broods. A natural circle!

Dwarf Crayfish or CPO, *Cambarellus patzcuarensis* Villalobos, 1943

The Dwarf Crayfish *Cambarellus patzcuarensis* can be socialised well with small fish. However, with certain fish species, such as the livebearer tooth carps, you have to reckon with them decimating the offspring of the crayfish.

This pretty **Dwarf Crayfish** originally comes from Lake Patzcuaro in Mexico and does not require excessively high keeping temperatures due to the high altitude of its native waters. It reaches sizes between three and five centimetres and in its orange breeding form is a pretty splash of colour for smaller aquaria. Unfortunately, a fundamental shortcoming in keeping this species lies in the breeding selection, as almost all orange animals, known as **CPO (Acronym for *Cambarellus patzcuarensis* orange)** in aquarium lingo, are said to come from very few original animals, which Brian Kabbes sorted out from the brown wild-caught animals and bred on. This is why there are always dramatic losses of offspring here, as the animals can obviously have genetic defects. On the other hand, there are also some very successful breeders, who have managed to compensate for these basic deficiencies by feeding the animals a healthy diet of catappa-leaves from the ***Terminalia catappa*** tree and keeping them in favourable conditions. Personally, I have not had any luck with breeding this species so far, but it is my conviction, that the brown wild form should be crossed into the existing breeding strains in order to prevent possible genetic deficits. Due to its small size and peacefulness, the CPO is one of the most suitable crayfish for aquaristics. It can be excellently socialised with small fish, such as **Neon Tetras** or **Guppies**, but not with freshwater shrimps, as they may fall victim to it. Unfortunately, the CPO only lives for one to two years, but under optimal living conditions it usually provides for sufficient offspring during this time, whereby 20 to 50 juveniles animals can be expected. The CPO does not attack the aquatic plants and can even help to eliminate a snail plague in the aquarium. This useful crayfish species should definitely be preserved for aquaristics through offspring breeding, but breeders should consider, how to increase the genetic potential of these animals in the future. Fortunately, various colour morphs of this species are already in circulation, so that the species can already be counted among the established species of aquarium enthusiasts. This crayfish is the typical example of successful marketing and establishment of a species in freshwater aquaristics without endangering the native stocks in Mexico by excessive natural removals. This is due to the fact, that the orange morph of this crayfish was first bred in Europe. Whether these crayfish are better bred in hard or soft water is a matter of debate, but reputable breeders give indications, that breeding in hard water of 6° German total hardness and above is more successful than in soft water. Basically, it should be pointed out, that the water chemistry has a direct influence on the colouration and well-being of this crayfish. Therefore, it can happen, that crayfish change colour in the tank of the aquarist after purchase from the dealer. Therefore, conscientious breeders/dealers should always communicate their own water parameters with the customer.

This artificial root made of polyresin has the unbeatable advantage, that its use neither influences the water chemistry nor colours the water brown.

Here, real humic acid-rich roots and calcareous marble slabs have been combined, resulting in a pH-neutral environment.

Infraorder *Caridea* – True Shrimps

The infraorder ***Caridea*** is the currently valid term for the **True Shrimps**; the original designation ***Natantia**** has since been marked as invalid. The true shrimps also belong to the order *Decapoda*, the decapod crustaceans, as they have a total of five pairs of legs. A common feature of all shrimp species belonging to the *Caridea* are the lamellar gills; these make it easy to distinguish them from the prawns belonging to the suborder ***Dendrobranchiata***, the treelike gills. In English-speaking countries, the members of the *Caridea* are usually named "shrimp", while the *Dendrobranchiata* are called "prawn". However, this has nothing to do with the scientific systematics at all and is constantly confused in the common language use. In aquaristics, various species of the *Caridea* have been bred for many years, and some of them are even traded at extremely high prices. In Japan, for example, there are said to be dwarf shrimps of the genus *Neocaridina*, which are traded by their breeders at prices of about 9,000 € for just one specimen. This deals with the fact, that the Japanese have always been keen on to breed animals in their national colours of red and white, which ideally might even bear the Japanese flag on their ***Cephalothorax***. Whether it makes sense to select animals in this way for commercial reasons alone is a matter for the reader to decide himself.

*** Lat. *Natantia* = "the swimming ones". Picture above: A shrimp from Scotland preserved as a fossil. It resembles recent (today living) species in all details.**

Amano Shrimp – *Caridina multidentata* Stimpson, 1860

This Amano Shrimp feeds on black beard algae right now.

Female Amano Shrimp, about 40 millimetres long.

This species, which is frequently offered for freshwater aquaristics, originally comes from Japan, Taiwan and Madagascar. After scientific revisions, this species, once traded as ***Caridina japonica*** **De Man, 1892,** has now been recognised as synonymous with ***Caridina multidentata*****,** described by **Stimpson** as early as **1860**. Which is why the older name has now been accepted as correct. Females reach sizes up to almost 50 millimetres total length and are thus clearly larger than the males. Unfortunately, it is unclear whether this species is hermaphroditic, in which the males later give rise to females, as is known from other shrimp species. This species is therefore a very valuable aquarium shrimp, because it actually eats the black beard algae, which often occur in freshwater aquaria and can ruin the plants of an aquarium by covering them with unsightly brown-black tufts. In addition, this shrimp lives up to six years, which is an almost biblical age for such a small aquatic animal. The **"Amano Shrimp"** owes its name to the professional Japanese aquascaper Takashi Amano, who was very fond of establishing this species in his landscape tanks for plant care. From Japan, it then began its triumphal procession in Europe. The Amano Shrimp carries very small eggs, from which tiny ***Zoea***-larvae develop, which can be reared in a brackish water tank at a salt density of 1.017. For this you need rearing aquaria. This means aquaria with thin films of green algae on the panes, which the juveniles feed on. The Logemann-brothers in particular have rendered outstanding services to the productive breeding of the Amano Shrimp in Germany. The females of this shrimp have conspicuously large abdominal shields. When they are carrying eggs, the fry can be seen shimmering through them, if you look closely. Once you have prepared a suitable rearing tank, you can immediately transfer the shrimp to the brackish water without any time-consuming acclimatisation, as they are not sensitive to oscillations in salinity. This is, because these shrimps also undertake spawning migrations from rivers and streams towards the ocean in nature and, as a fully developed young shrimp in the sea, naturally also vice versa. Once you have succeeded in raising fully transformed young shrimp in brackish water, they can be switched back to fresh water after a careful acclimatisation. Amano Shrimps can be socialised with non-predatory small fish without any problems. They do not show any shyness after a period of acclimatisation. However, when placed with the wrong fish, they take refuge in hiding places and then lead an inconspicuous nocturnal life. A new approach to keeping and reproducing this shrimp could be to socialise them with small brackish water fish and thus breed them permanently. Conceivable species for this would be the small **Bumble Bee Gobies (*****Brachygobius spp.*****)**, but perhaps also little **Endler-Guppies (*****Poecilia wingei*****)** or similar species. In such an approach, the majority of the shrimp larvae would serve as food for the fish. But on the other side this could also increase their spawning readiness and productivity.

Red Fire or Red Cherry Shrimp or "Sakura" – *Neocaridina davidi* Liang, 2002

Such deep red coloured animals are also traded at high prices as "Sakura". In Japan, however, this is actually a term for edible shrimps living in the sea.

The shrimp community of the basin most probably consists of only one species, which is, however, very variable with regard to its colouration. Especially in the beige-brownish coloured specimens the dorsal stripe is clearly visible.

This shrimp, which is also called **Red Cherry, Fire** or **Backstroke Shrimp**, was also exported to Germany towards the end of the 1990s. It originally comes from China and Taiwan, from where it was further exported and can now even be found in Hawaii. It can reach a length of about 30 millimetres in the male sex and about 40 millimetres in the female. The species is extremely variable in colour, but the dorsal stripe can be seen in almost all colour variants. The Fire Shrimp can also be kept and bred in unheated aquaria. It needs organic waste and fine fish food for nutrition. If it feels comfortable in an aquarium environment with a lot of cover, it can become very productive. This is, why it has been bred as a fish food originally. You can easily socialise this shrimp with small fish, but you should always get groups of at least ten or more, to make them feel comfortable. They thrive best in tanks with detritus corners that are not sterilely cleaned. Here their juveniles will also find enough infusoria and the like to feed on. This shrimp can even reproduce in aquaria with viviparous tooth carps, as long as it finds enough hiding places and dense planting. It is also one of the shrimp species, which help to contain and efficiently combat black beard algae infestations. These shrimps are excellent for socialising with crayfish, which is described in more detail in the final section of this book. The shrimps can regularly be observed not only eating the food remains of the crayfish, but also eating the collagen secreted during the moulting of a crayfish, with the help of which the crayfish slides out of its old exoskeleton. So this species belongs to the most useful crustaceans for any freshwater aquaristics.

Crystal Red Shrimp, *Caridina cf. cantonensis* Yü, 1938

This tiny shrimp has been exported via Hong Kong for many years, but it seems reasonable to assume that the species is also present in mainland China. It reaches a total length of about 20 millimetres and lives only one to two years. It is available in numerous colour varieties, ranging from reddish brown, bluish, transparent, reddish, beige, striped, dotted or spotted. The species has only a short rostrum, but produces relatively large eggs that reach almost one millimetre in diameter. Especially the cultivated forms of this shrimp, which are traded as **Crystal Red Shrimp**, can react sensitively to unsuitable water conditions. If the environment does not suit them, they may not be able to reproduce. If no effort is made with the water chemistry during acclimatisation, they may even fail completely. This is not always due to poor water quality, but can also be related to bacteriological conditions in the water, which unfortunately cannot always be influenced by the breeder. For example, a well-known importer confided in me, that it sometimes happens with diverse fish or shrimps, that whole consignments die because of such problems. And that then de facto nothing can be done about it. This is comparable to the intestinal *bacteria* of humans, which go crazy when eating exotic food, because they cannot cope with the germs it contains, while the same food does not bother local people ("Montezuma's revenge").

Bamboo Shrimp – *Atyopsis moluccensis* De Haan, 1849

Here a Bamboo Shrimp on a mangrove root. These shrimps like to crawl up exposed wood, plant stems or pipes in order to reach more flow-rich areas in the aquarium. It is therefore especially important to tape off covers and corners near such installations, otherwise the Bamboo Shrimp could escape from the tank.
The specimen shown here has a light-coloured dorsal stripe along the entire body length and could thus have been imported from the Philippines.

The claws of the Bamboo Shrimp, shaped as fine fans, are an adaptation of many marine crustacean species to a filter-feeding diet.

A typical behaviour is to seek out places with strong currents.

The **Bamboo Shrimp** is also called **Radar Shrimp** or **Moluccan Mountain Shrimp**. It is widespread in the coastal regions of Southeast Asia and can be found in small streams and rivers with a connection to the sea. This is especially the case on the Indian subcontinent, on the coasts of Thailand, the Philippines, the Moluccas and numerous other islands in this Indopacific areas. The Bamboo Shrimp reaches a total length of almost 90 millimetres and has a very variable colouration with various beige and brownish patterns and stripes. These patterns can vary depending on the location, which is certainly due to the fact that different localities also produce different habitats, to which the animals then adapt with their colouration. The different species of this family can best be distinguished by their rostral spines, for which it is best to examine an ***Exuvium*** using a magnifying glass or microscope. The adult Bamboo Shrimp can be found in nature under stones or on wood, where they usually prefer sections with a lot of current. The juveniles, on the other hand, stay in calmer areas. Contrary to the name **"Mountain Shrimp"**, this species does not penetrate into higher altitudes near the coast, but can still be found at distances of up to 70 kilometres from the coast. For reproduction, they absolutely need salt water, so that their breeding in aquaria has not been known so far. Presumably, their larvae require a salinity of 1.030 or more, which unfortunately is not yet known exactly. However, this would be supported by the large distribution area of this species, as the Bamboo Shrimp obviously uses the Indian Ocean as a distributor for its larvae. It is also likely, that their larvae live for a similar length of time as the larvae of pure marine shrimps in the sea, so that their development period may even be up to a year or more. For this reason, only adults are caught and exported for aquaristics, which is said to have already brought them to the brink of extinction in some areas. Therefore, it would be very desirable, if dedicated aquarium enthusiasts would seriously take up the breeding of this species in order to contribute to the avoidance of imports. In the aquarium, these shrimps can be kept well and for long terms, as long as they are not socialised with fish, which chase the shrimps or drive them out of their territory. It is a proven fact, that they can be kept for up to twelve(!) years, which is an enormous age for an invertebrate of this size. The Bamboo Shrimps feed on all kinds of food remains and detritus, which is why you do not necessarily have to feed them specifically. They also need wood, stones, hiding places and structured back walls, which they can then climb. However, you have to be very careful to secure corners and pipes with tape, so that the animals cannot leave the aquarium.... And of course this is their master discipline, which can drive an average aquarist becoming crazy.

Superfamily *Palaemonoidea* – Glass Prawns

From a purely systematic point of view, the large-armed shrimps of the genus ***Macrobrachium*** belong to the ***Palaemonidae*** family. While our native shrimps of the genus ***Palaemon*** already have species, that can survive in brackish or fresh water, this characteristic has already been developed to such an extent in the large-armed shrimps, that some species can only be found in pure fresh water. As a rule, these are also relatively easy to breed. Most species of the genus ***Macrobrachium***, however, live in coastal rivers or areas near the sea-shores. Here they entrust their fry to currents, which carry them into the ocean, where they develop through various larval stages into adult shrimps. Since some species are also commercially interesting for humans. And because of their excellent meat, they are also bred in corresponding aquacultures. Such large-armed shrimps often exhibit a striking sexual dimorphism, in which males have considerably elongated claws, that can become twice as long as those of females. Larger species are often skilled fish hunters or scavengers that usually have to be kept alone in the aquarium, because they behave very aggressively towards conspecifics. Smaller species, on the other hand, can sometimes be socialised in freshwater aquaria with fish and other animals that are not too large.

Female of a *Macrobrachium species* from aquaristics. This is a species with typical sexual dimorphism, in which the males have considerably longer claws than females.

Red Claw Shrimp, *Macrobrachium dayanum* (Henderson, 1893)

This extremely pretty shrimp originates from the freshwaters of India and Bangladesh. Males can reach a body length of at least six centimetres under optimal care conditions in large aquaria, the females are usually somewhat smaller. The **Red Claw Shrimp** only shows a red-black annular pattern on its claws when juvenile, which then gives way to a uniform reddish-brown shade in adults. This shrimp is a very territorial species in which a male defends a certain territory against sexual mates, which can lead to the death of the inferior males in aquaria, which do not offer enough living-space. So it is best to keep a male together with a harem of several females. The animals are very productive and can reproduce again about every two months. This species is very valuable for aquarists, because it likes to eat snails and does not attack fish and plants. However, breeding forms or fish with veil fins are the exception. It is not advisable to keep Red Claw Shrimps together with crayfish, because they can attack small crayfish at the latest during moulting. Keeping them together with large species such as the **Australian Red Claw *Cherax quadricarinatus***, on the other hand, can make sense, if you want to limit the productivity of the crayfish and accept, that the shrimps may prey on parts of the crayfish fry. The aggressions of the Red Claw Shrimp may be limited by a poor protein nutrition. And furthermore by a richly decorated aquarium with wood, stones and many useful structures, which invite the animals to climbing. However, the tank should be well covered, so that the animals cannot escape.

Indopacific Glass Prawn, *Macrobrachium idae* (Heller, 1862)

Male Indopacific Glass Prawn; typical are the extremely elongated claws, which represent a clear sexual dimorphism.

Note the six extremely long antennae of this species, with which it scans its entire environment and locates prey and enemies. These antennae are irreplaceable for orientation in muddy waters. Experiments proved, that shrimps without eyes have a significantly higher chance of survival in nature than shrimps without antennae!

The rostrum can be used to distinguish this species from other Macrobrachium species of this faunal circle.

This extremely large-armed shrimp, which reaches truly exorbitant sizes, is found in the coastal waters of the Indian Ocean. Its range extends from Australia to Southeast Asia and East Africa. Males reach a body length without claws of up to about 11 centimetres; body and claws of the males may reach a total length of about 40 centimetres and more. Females have shorter claws and are somewhat smaller and stockier. In addition, their anterior abdominal shields at the beginning of the ***Pleon*** are larger than those of males. As adult, ***Macrobrachium idae*** inhabits river systems and estuaries connected to the sea. The animals also have a certain brackish water tolerance. They then entrust their larvae to the current, which transports them to the ocean, where they develop via various ***Zoea*** larval stages into adult shrimps, which then return to the rivers. In the aquarium, this shrimp unfortunately proved not to be a very persistent species, and when acquired as an adult it only lasted a few weeks. This large-armed shrimp must be kept without any fish stocking, as these are preyed upon by it. Furthermore, keeping this species in brackish water should be beneficial for a longer life. I myself received the species from an importer, who considered it to be a subspecies of the **Giant River Prawn** ***Macrobrachium rosenbergii***. If that is true, may be checked and researched by the molecular biologists someday. Such prawns are interesting research objects to biological studies anyway, so that keeping and hatching them means a serious challenge for any true prawn-enthusiasts.

Giant River Prawn, *Macrobrachium rosenbergii* (De Man, 1879)

Semi-adult Giant River Prawn, about 8 centimetres long.

***Exuvium* of a Giant River Prawn.**

In the past, the **Giant River Prawn** probably fell victim to confusion with another species after its first description, so that the exact species status must currently be considered "pending". However, there are indications that, for practical reasons, the "wrong" taxonomic species name will continue to be declared correct. The Giant River Prawn is probably the biggest known large-armed shrimp, because without claws, males can reach a length of up to 32 centimetres, and females a length of up to 25 centimetres. Adult

males can have claws, that reach more than 50 centimetres in length! The claws of females remain considerably shorter, which is a typical sexual dimorphism. Another special feature is typical for the Giant River Prawn, and can also be easily noticed in juveniles to distinguish them from other similar species: On the middle of the ***Abdomen*** they bear a row of orange spots, one on each side of the body, with each spot sitting exactly between two abdominal segments. The Giant River Prawn was originally found in various Southeast Asian countries, preferring muddy estuaries and rivers with a connection to the sea. The Mekong Delta, in particular, is an ideal habitat for the Giant River Prawn. Here, due to the low-lying river delta, a tongue of saltwater pushes up to 150 kilometres inland, because the saltwater from the sea is heavier due to its higher density than the freshwater of the Mekong, which drains into the ocean via this tongue. Since freshwater and seawater hardly mix, the phenomenon occurs at such places, that marine animals - such as bottom-dwelling rays - can penetrate far inland, although they do not actually tolerate freshwater. It is in this area of tension that the Giant River Prawn has settled. This place is so ideal for this species, because the larvae of the Giant River Prawn need sea water for their larval development. Thus, the pregnant females do not need to migrate all the way to the sea to transport their larvae into the salty element and can penetrate very far inland themselves. The brood then passes through ten different larval stages until it transforms into the small post larval developmental stage. Depending on the temperature, the larvae of the Giant River Prawn develop within only 32 to 36 days, until they finally reach the post larval stage with a size of about 8 millimetres. Due to this short development time, commercial shrimp farming is very profitable, so that the Giant River Prawn is now also farmed in Israel and the USA. Giant River Prawns are territorial animals, in which a male defends his territory against other males and protects the females in his territory until they have moulted. Then the males' mate with the females, and after a short time the females eject several thousand eggs, which are carried under the ***Abdomen*** until the brood hatches. Unfortunately, the territorial requirements of Giant River Prawns are so extreme that they have to be kept individually in the aquarium. It is also not possible to keep pairs permanently, because sooner or later the animals will damage each other until the inferior animal dies. Giant River Prawns cannot be socialised with fish or crayfish either, as they will sooner or later attack even large and defensive animals. What a Giant River Prawn then clutches with its claws, it will not let go of again - it is more likely to have its claw arm torn off. Because its clawed fingers lock, after the prawn has grabbed it like a door that has fallen shut. The local fishermen in Southeast-Asia take advantage of this, and so they catch these prawns from small wooden boats with a primitive fishing rod. Once the shrimp is hooked on the bait, the fisherman can easily lift it out of the water together with the bait and hoist it into his boat. Nowadays, animals caught in this way hardly ever make it into the international fish trade, because the market here is dominated by farm shrimps. On the one hand, it is good, that natural resources can be conserved through artificial reproduction. But on the other hand, it is problematic, because mangrove forests, which provide natural coastal protection, are destroyed for shrimp farms. In addition, farmed prawns are often pumped full of antibiotics, because factory farming leads to epidemics and diseases. And very often, marine resources of all kinds are consumed to raise farmed prawns, which are then lacking elsewhere and can lead to a decline in fish stocks. Aquaculture in regions where Giant River Prawns do not actually occur regularly, also leads to escaped animals endangering and displacing native other species. Keeping Giant River Prawns in the aquarium is quite exciting. The young animals grow up extremely fast, and within only two to three months a juvenile of 6 centimetres length can have doubled its body length! However, as they can usually only be kept in a species tank, dealing with this species is reserved for specialists and enthusiasts.

Giant River Prawn and Mangrove Crab inhabit different zones of the same habitat. This is why they can be socialised quite well in tanks with land access for the crabs.

Baltic Prawn, *Palaemon adspersus* Rathke, 1837

The **Baltic Prawn** is widespread, as it is regularly found in the Black Sea, the Mediterranean, the Atlantic, the southern North Sea, around the British Isles, in the Baltic Sea and as far north as Norway. It has now been further spread by shipping even to the coasts of Africa and India. It has two to six rostral spines on the top of the rostrum, while there are only two to four teeth below. Like some other species of this genus, it also has the typical blue-yellow ringed legs and scissor legs. Their body colouration, however, can be very variable, because depending on location and season, one can find monochrome brownish or beige-coloured animals, as well as bright yellow or even slightly orange-tinged specimens. The Baltic Prawn also tolerates low salinity and higher temperatures, which is why it is easy to keep in a brackish water aquarium. As they can often be caught in large numbers with a landing net on groynes and tide poles, they are also excellent live food for larger predatory fish. Unfortunately, shrimp species introduced into the Baltic Sea in recent years, as well as fish from other parts of the world's oceans, are on the rise, so that the occurrence of the actually common Baltic Prawn may be declining there. It is also unclear at present whether and with which introduced shrimp species the Baltic Prawn could mate. Unfortunately, the resulting bastards can hardly be defined as belonging to any species.

Estuary Prawn, *Palaemon varians* (Leach, 1813)

Estuary Prawn – female with eggs.

The **Estuary Prawn** is also found from shallow water onwards. This prawn, which can grow up to 5 centimetres, is difficult to spot because it can take on the colouration of the respective substrate. It can take on colours ranging from translucent to brown and green. This prawn is very often found in drainage ditches or harbours with low salinity. It also likes to settle on wooden piles with algae or mussel growth. It is often traded and shipped alive as fish-diet. According to my aquaristics experience, it can be kept for several months in a pure freshwater aquarium at room temperature after appropriate acclimatisation.

Grass or Rockpool Prawn, *Palaemon elegans* Rathke, 1837

In warm temperatures this prawn gets small yellow speckles on the whole body and the rostrum glows "golden".

The **Grass** or **Rockpool Prawn** is a very common prawn that can be found on the North Sea and Baltic Sea, as well as in the Mediterranean. It has probably been spread halfway around the globe by ships, in whose ballast water its larvae travel. There is a lot of confusion in the literature regarding its Latin name, as it has been repeatedly misnamed ***Leander squilla, Palaemon squilla***, etc... Rockpool Prawns usually grow to about 6 centimetres in length and prefer to sit on sheet piling, on stones, on piers and near locks and on the quay walls or harbour piers. They are small, agile omnivores, which feed on anything that seems edible to them. These shrimps can be kept excellently in the aquarium, tolerating even tropical temperatures and still bearing eggs regularly at temperatures of about 25° Celsius. The females can be recognised from a size of about 4-5 centimetres by the larger abdominal shields, the males by the smaller shields. The females release free-swimming ***Zoea*** larvae which, after a planktonic phase of about 3 weeks, transform into finished small prawns and then orient themselves towards the bottom. They have already been propagated in the aquarium. If you take a closer look at these animals, you will make some amazing discoveries: Their claws are ringed in blue and yellow, and the movable claw finger is located at the bottom of the claw, and not at the top as in most other crustaceans. When the water is particularly warm, they develop small yellow spots that cover their entire body. The inside of their frontal spine, called the rostrum, seems to glow like a golden line. If you hold your hand in the water, they will even come up to your hand and search it for edible particles, probably small skin flakes. There are several species of rock or grass prawns, which can essentially be distinguished by their rostral spines, their origin and also their colouration. Some confusion has unfortunately arisen, because some species have been described several times by scientists as new species in each case, and in the meantime new species of the genus *Palaemon* have also been introduced to Europe by ships. Since these species all look very similar, only rostral spine counting, measurement of body proportions or molecular genetic analysis remain for a really exact species identification. The Rockpool Prawn was included in the canon of this book, because it can be found on the coasts of the Baltic Sea even at very low salinity levels. However, this should not obscure the fact, that it is a predominantly marine species. Therefore, it can only be kept permanently in a brackish or seawater aquarium. It also tolerates room temperature and provides offspring all year round. The larvae can be reared with appropriate plankton substitute food, whereby the development period depends on the temperature. The first small offspring prawns can be expected after about two to three weeks with optimal larval food, as they go through relatively few larval stages and then quickly become benthic. Breeding such prawns makes sense, especially if you need them or their offspring as diet for breeding other animals, such as seahorses, for example.

Infraorder *Brachyura* – True Crabs

The **Short-Tailed Crustaceans** carry their ***Abdomen*** folded under their body. They also have ten pairs of legs, the first pair bears claws. This group includes very different representatives such as the **Land Crabs, Spider Crabs** and the pelagically **Swimming Crabs** on the other. In the latter, the last pair of legs is flattened, so that they can swim with it in the open sea. Most crabs are generalists, eating anything they can catch. They often eat carrion as well, taking on the important role of health police for their particular habitat. Some species have plenty of usable meat and are commercially important to local fisheries, such as the large species of swimming crabs, which also enter estuaries from the sea in subtropical and tropical coastal waters. In the various ecosystems of our waters, many crab species are an important link in the food chain. For example, crabs in particular make up one of the main food components of many birds and fish. Especially their larvae mean an important addition into the marine food-chains. This is also necessary in the ecosystems of the coastal waters of the world's oceans, because otherwise the number of crabs would rapidly overgrow and too high a population density could have dramatic consequences. Due to the international movement of human goods, various crab-species are constantly being redistributed around the globe, some of which have a significant impact on native fauna. In addition, crab-species, such as the **Chinese Mitten Crab**, also directly harm humans themselves by massacring fish in fish traps, damaging fish fry, and puncturing dams and dikes. In sum, such displaced species often cause millions of dollars in order to repair the ecological damage that has occurred. Some damage is irreparable. Most crab-species are relatively easy to keep in an aquarium, as long as you know what their living requirements are. In addition to appropriately large tanks, however, you also need a very solidly set up aquarium in which the animals cannot cause any stones to collapse. Otherwise there is a risk of domestic flood catastrophes if "only" the bottom pane is cracked due to the animal's activity... If such a mishap does occur, you have to weigh up whether to try to seal the leak under water with special silicone and a small pane of glass in the case of a minor leak, or whether to empty the tank completely. Don't worry - silicone also hardens under water. And if there is still a tiny crack afterwards, you can try to seal it with calcium powder, which can be successful after a few days. In any case, one should remain calm and carefully consider one's own course of action. Panicky and hasty decisions in such situations usually lead to the loss of the animals. Another reason to lose aquarium inhabitants is, if they manage to escape from the tank. Especially **Mangrove-** and **Mitten Crabs** are experts in breaking out. They are even able, to lift up cover panes. In some cases, it is better to weigh them down with bricks or to put a metal grid of rabbit wire on the tank. Otherwise, the crabs will walk around in the living room...

Family *Gecarcinidae* – Land Crabs

Land Crabs are related to the **Talon Crabs** of the family ***Grapsidae*** and, like them, have been positioned in the superfamily ***Grapsoidea***. In contrast to the *Grapsidae*, they usually have two unequally sized claws and their physique is much bulkier and taller. They are native to the coastal zones of the tropics worldwide and live semi-aquatic or terrestrially, staying in the forests and swamps near the coasts. They depend on proximity to the sea, as they must entrust their brood to seawater for further development. On some South Sea islands, such as Christmas Island, there are large annual spawning migrations of female crabs to the sea. And then some time later the reverse invasion of young crabs onto the island, which is a very impressive natural spectacle because of the large numbers of crabs. Land Crabs are nocturnal and feed on carrion, fruits and other organic food. They are also eaten locally and caught using special traps. Alive, they are a food that can thus be kept fresh even in tropical areas for several days. Sometimes they are only to be enjoyed with caution, as they can eat poisonous plants, which then make the Land Crabs inedible. Furthermore they could be infected with parasites, which can cause serious diseases! Therefore, they must be kept for a few days and fed with harmless plants and fruits before being eaten. For aquaristics, the beautifully coloured species ***Cardisoma armatum*** is regularly imported from West Africa, but other species of the genus ***Cardisoma*** are also regularly imported for hobby purposes. While ***Cardisoma armatum*** is still a semi-aquatic species, that can be kept well in a heated aquaterrarium of appropriate size, other species of this genus can usually only be kept for a longer period of time in a terrarium with high layers of soil and water bowls. Unfortunately, they are not often seen, because they hide during the day due to their nocturnal lifestyle. There are anecdotes of sailors, who brought exorbitantly large crabs home alive from the Caribbean as a souvenir, and then kept them free-ranging in their homes. This may indeed have worked for a while under summer conditions, but it cannot be considered species-appropriate husbandry, as the crabs can catch cold!

Mouthless Land Crab, *Cardisoma crassum* Smith, 1870

With a bit of luck, you can "see" *Cardisoma crassum* in the terrarium...

The **Mouthless Land Crab** originates from Central and South America and is also found in the Gulf of California, especially in Mexico. With a leg span of more than ten centimetres, it already reaches quite impressive sizes and must therefore be kept in correspondingly large terraria. Here it needs an arrangement of branches for climbing, as well as a high layer of sand or earth for digging; in addition, a water container should not be missing. Such a terrarium can hardly be planted, as the crabs eat or destroy plants. It is also difficult to keep them together with other animals, as these can easily become their prey. To do their moulting business they burrow into the ground like Land Hermit Crabs, so that they grow unnoticed by the keeper. They can be fed with all kinds of fruits, but you should try out the whole range of available fruits and vegetables, as land crabs can sometimes have individual preferences. You have to be a real enthusiast to take care of such nocturnal crabs, because you rarely see them during the day. Breeding them is also unlikely to be possible due to their marine reproductive behaviour, not to mention the problem of keeping them in pairs. Nevertheless, it is interesting to deal with such unusual creatures from time to time.

Here is a male of the Mouthless Land Crab from above and below. The sexual dimorphism of this crab is particularly striking, with the males having larger claws in this case. This specimen probably lost a claw, which is why it had to be regenerated and is now present again in a somewhat smaller form. This usually happens after a scuffle with conspecifics or after a failed moult.

Family *Grapsidae* – Talon, Shore or Marsh Crabs

Like the **Land Crabs** of the ***Gecarcinidae***, the members of this family also belong to the superfamily ***Grapsoidea***. However, it is very easy to distinguish them from the Land Crabs, because their claws are always the same size. They are mostly marine crabs, many of which live semiaquatically. This way of life is usually encouraged by the fact, that they live in tidal areas, where they wait under stones or buried in sand at low tide for the next flood. They are very nimble opportunists, eating anything they can somehow overpower. They are also excellent climbers, which makes it very difficult to keep them permanently in an aquarium. Some species of this family, such as the **Chinese Mitten Crab** introduced into German rivers, make extensive spawning migrations from the rivers to the sea, where they entrust their larvae to the North Sea for development. In doing so, dikes and weirs are no obstacles for them, as they can then climb over them in huge layers at certain times. In the process, they also cause considerable damage, as they can puncture dikes, which can then have fatal consequences for the people living there during stormy floods. But also by eating fish fry, the Mitten Crabs have become a major ecological problem, apart from the fact that they also like to steal anglers' bait from the hook and eat fish from fish traps. For this reason, such animals must never be released anywhere, because you never know what damage they might cause. Keeping members of this family in aquaria is easy, because such crabs are not choosy about their food. Unfortunately, because they also eat conspecifics after moulting. And they are masters at getting into the open through the smallest cracks and crevices. Mitten Crabs can be kept just as well in a freshwater aquarium as in a saltwater or brackish water tank. If you get tropical species from this family in the trade, you should enquire very carefully about their exact origin, so that you can house them reasonably species-appropriate in a tank, that should be as similar as possible to their native habitat. Unknown species of doubtful origin pose a great risk, as it may be difficult to distinguish whether they are a marine species or a freshwater species. On the one hand, such species offer a rich field of activity for pioneers, on the other hand, the risk of losing the animals through incorrect care is very high. But this is something the dedicated hobbyist must decide for himself.

Mitten Crab, *Eriocheir sinensis* H. Milne Edwards, 1853

The **Chinese Mitten Crab** was first caught in Germany in the Aller River in 1912. It was brought into our waters by ships and must therefore be counted among the species defined as neozoans. The Mitten crab originally comes from the Yangtze River in China, from where it was probably introduced in larval form with the ballast water of ships. As juveniles, they live in rivers and freshwaters, but in order to reproduce, they migrate back to the sea. In doing so, they migrate hundreds of kilometres upstream and also use ships to reach remote waters. In the meantime, they have even reached Lake Constance by the Alps. While they initially spread through the North Sea into northern Europe, they have now also colonised most of the larger inland waters far from the sea. Mitten Crabs are a sought-after delicacy - in China - and a serious threat to our native fauna in our rivers and lakes. Resourceful businesspeople have already started exporting Mitten Crabs to China, where they have become overfished. Moreover, Mitten Crabs dig through dams and dykes, posing a serious threat to people living in areas prone to storm surges. Even anglers hate them because they steal baits from their hooks, damage fish fry and plunder fish traps. In order to curb the crab plague somewhat, mechanical shredders are now installed at weirs and dams to chop up Mitten Crabs as they climb over the weirs. Keeping them in the aquarium is possible in principle, but it is a challenge for the keeper to house them escape-proof. They can lift carefree cover panes and escape through the smallest cracks and crevices without any problems. In my own cellar they escaped and even reached the neighbour's cellar! But I was lucky, because my neighbour was an enthusiastic aquarist, too.

Estuarine Mud Crab, *Rhithropanopeus harrisii* (Gould, 1841)

The **Estuarine Mud Crab** grows to about 25 millimetres wide and originates from the north western Atlantic. Through shipping it became widespread and now lives in the Northwest Pacific as well as in the Aral Sea, the Black Sea, the Caspian Sea, parts of the Adriatic Sea and the coasts of the North Sea and the Baltic Sea. At first glance, this crab resembles tropical **Mangrove Crabs** or small **Mitten Crabs**. And the square shape of the head breast makes it easy to see that they also belong to the superfamily ***Grapsoidea***, the **Talon-, Marsh-** and **Shore-Crabs**. The **Estuarine Crab** prefers rather low salinity and therefore feels particularly at home in the Baltic Sea. But it has also been known from the North-Baltic Sea Channel since some years. The male specimen shown here was caught from a wooden pole in the harbour at the Baltic Sea Station in Lübeck-Priwall, where it was sitting among mussels and barnacles. In the aquarium, this crab leads a very hidden existence, but is excellent and long-lived here. As it prefers brackish water habitats, it can only compete with our native **Green Crab *Carcinus maenas*** in these locations with low salinity. Unfortunately, it is rarely seen in the aquarium, as long as it is offered sufficient hiding places. This suggests that it belongs to the nocturnal species.

Superfamily *Potamoidea* – River Crabs

This superfamily contains very many genera and species that are distributed worldwide. The best known is probably the **European River Crab *Potamon fluviatilis***, which has even become a motif on ancient coins from southern Europe. River Crabs are pure freshwater dwellers that reproduce far from the sea. They have strongly toothed claws and often show no sexual dimorphism worth mentioning. They often live predatorily, eating anything they come across that they can overpower. They often even eat amphibians, but even inferior conspecifics can quickly become their prey. Therefore, the representatives of this family can often only be kept individually. The name **"River Crab"** is rather misleading in some cases, because they also occur in temporarily drying small water bodies in some locations, and some species live semi-aquatically. Therefore, most crabs of this family must be offered an aquaterrarium for permanent keeping. This should have many structures, be large enough and offer the animals many hiding places. Many species are nocturnal, so that the keeper hardly ever sees them during the day. Wild-caught crabs are sometimes heavily contaminated with gill parasites and other diseases, so that quarantine for new imports makes sense. Especially empty moults or even dead specimens of a new import should always be checked for parasites. In some cases, treating infected animals with a brackish water bath can be successful. River crabs document the completion of a line of development of crabs from the sea to the land, which is why they have been included here in short form in the context of this book. Since they carry their brood until it is fully developed, breeding of many species is possible. In recent years, there has been a steady increase in the number of aquarium enthusiasts who have devoted themselves to breeding these species, which are often also attractively colourful. The care requirements of the individual species can be very different, so that dealing with these animals is rather something for advanced aquarists and pioneers.

Siam Crab, *Thaipotamon siamense* A. Milne-Edwards, 1869

The **Siam Crab** documents the completed transition from the sea to freshwater, as it is already no longer dependent on salt or brackish water. So it belongs to the so-called secondary freshwater fauna. The Siam Crab can reach a leg span of about ten centimetres, what means, that it should be kept in tanks with a length of 80 centimetres and more. Internally, this crab is relatively peaceful and can therefore also be kept in small groups. It should be kept in a spacious tank with a land part, which must be appropriately secured against the crabs breaking out. It is advisable to set it up with wood, fixed stones, sand and foliage to somewhat imitate the natural habitat. The crabs usually do not eat fish and do not prey on them. They should be fed with fruits, possibly also with protein-containing granulated food. This crab no longer needs the sea to reproduce, but is a pure freshwater inhabitant that also carries its young here. Whether breeding in the aquarium has already been successful is unfortunately not known to me personally, but it would be conceivable. Therefore, the incentive to keep such animals is considerably higher than that of Mangrove- or Land Crabs, which are usually difficult or impossible to breed.

Family *Sesarmidae* – Mangrove Crabs

The **Mangrove Crab**-family also belongs systematically to the superfamily ***Grapsoidea***, and is therefore closely related to the **Land Crabs** of the ***Gecarcinidae*** and the **Talon Crabs** of the ***Grapsidae***. In Mangrove Crabs, both claws are usually equally developed, but they show a classic sexual dimorphism, as the claws of males are stronger and more developed than those of females. They are found everywhere in the tropics where dense mangrove stands line the coasts. They live not only on the stilt roots of the mangroves, but also on land and in the adjacent mudflats. The different species of Mangrove Crabs have conquered the most varied habitats and regions for themselves. Just like the mangroves, the Mangrove Crabs tolerate strong fluctuations in the salinity of the water entering this habitat. Just like the **Land Crabs** of the ***Gecarcinidae***, Mangrove Crabs entrust their larvae to the currents for development, which then carry the larvae into the sea. Here they develop over several developmental stages into the finished small crab. Mangrove Crabs can now also be bred, but it is very time-consuming to raise the larvae in special incubators with the appropriate plankton food. Mangrove Crabs can be kept very well and for a long time in aquaria with wooden roots, through which they can also leave the water, although groups of these animals should always be kept as small as possible. Nevertheless, an aquarium should not be too small for these animals, as they also occasionally engage in territorial disputes, which can then end with the loss of claws, other limbs or even the loss of animals. Small species of this family can be socialised with fish without problems, as they do not prey on live fish. However, as scavengers and health police, they can sometimes feast on sick or dead fish. If fed well, they usually do not attack the plants in the aquarium.

Mangrove Crab– *Sesarma bidens* De Haan, 1835

A Mangrove Crab lost a claw in a scuffle.

Mangrove Crabs are very fond of climbing on wooden roots.

This small **Mangrove Crab** is widespread along the coasts of Southeast Asia and is regularly imported for aquaristics. Whether this is actually ***Sesarma bidens*** can unfortunately only be determined with very specialised literature, the study of which would then also take a correspondingly long time. Therefore, the species shown here can only be classified as ***Sesarma bidens*** with reservations, as very probably other species are also traded under this name. Also, these crabs have already been traded as **"*Pseudosesarma moeschi*"**, which cannot be entirely correct either, as there is no species of the genus *Pseudosesarma* with the species name "*moeschi*", but only one species of the genus *Sesarma* with the species name *moeschi*. To what extent half-knowledge or wishful thinking played a role in the choice of the traded species name, and where the error crept in, can unfortunately no longer be traced. The Mangrove Crab can be kept well in aquaria with a tank length of about one metre or more, as long as the animals are given the opportunity to go for a walk in the fresh air. However, escape routes must be secured with adhesive tape via cables and hoses, otherwise the Mangrove Crabs will take long walks in the keeper's home. This can be quite scary, when they look like a big grey spider at first sight, covered in all kinds of fluff - cries of fear from the housewife are guaranteed! The best way to keep these animals would be in an aquaterrarium with a clearly defined water and land section. The transitions can then be designed with stones or wood. It would be very attractive to design a biotope aquarium with brackish water in which, for example, **Archerfish** and other species such as the brackish water preferring **Corsula Mullet** ***Rhinomugil corsula*** (pictured above), which have a certain tolerance for salty water, could be kept. Going beyond this, of course, one could also set up a tank where tidal effects cause the level of the water to fluctuate. Here, one could even try to socialise Mangrove Crabs and small **Mudskippers**, but the latter must not be too large, as large Mudskippers could otherwise become dangerous to the Mangrove Crabs. The creativity of the keeper is ultimately only limited by the space and financial resources required. However, recreating habitats is ultimately a much nicer approach to professional aquaristics than disjointedly socialising and exhibiting animals.

Tree Climbing Crab – *Episesarma mederi* (Milne-Edwards, 1854)

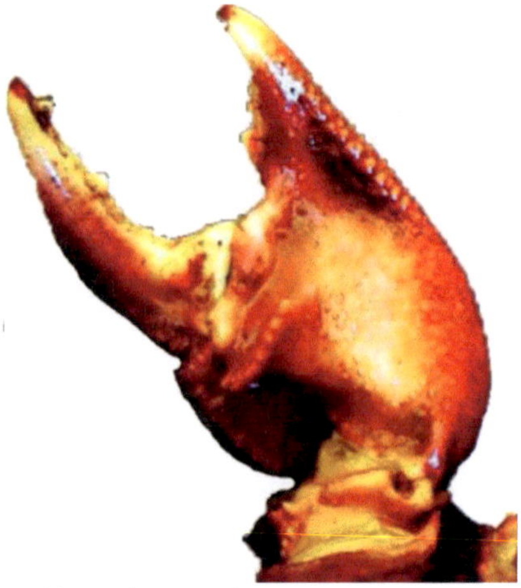

Male from below. **Claw in detail.**

This species originates from Southeast Asia and Thailand in particular. It reaches the considerable carapace length of about 4 centimetres and thus belongs to the larger species of its family. In Asia, the species has so far only been used as food, but it is also conceivable that it will one day be imported alive for aquarium enthusiasts. I found the specimen pictured here in an Asia shop, where it was offered in frozen form for Asian gourmets.

Family *Portunidae* – Swimming Crabs

This family of crabs is mainly distinguished by the strongly flattened paddle-feet, which are the last pair of legs at the back of the body. **Swimming Crabs** are found worldwide in all seas and have colonised all habitats from lagoons and shore zones down to the abyssal deep sea. Even the brackish waters of estuaries are regularly invaded by some species. Many species of this family live free-swimming in the water column of the oceans, where they pursue a bathypelagic lifestyle. They are usually strong predators, hunting small fish and other small prey in large schools, and in turn serving as prey for larger fish. Swimming Crabs have also been known to scare and to fool the captains of ships, as their shells reflect the echosounder waves of ships and thus simulate dangerous shallows in places, where the sea should be several thousand metres deep according to the nautical charts. Freshly moulted Swimming Crabs are also often victims of their own kind; moreover, in this state they are also often marketed for human consumption. For this purpose, they are kept in special cages and the freshly moulted animals are sorted out before their fellows do eat them. Some of the deep-sea species are marketed as delicacies at high prices. They are then traded as **"Blue Crab"** or **"Flower Crab"**, but these names often conceal different species, which are then usually traded as ***"Portunus pelagicus"***. Keeping Swimming Crabs in aquaria is difficult, because they require a lot of oxygen and become dangerous to their own species first, when they have moulted. Furthermore, once they reach a certain size, they can no longer be kept together with other animals without hesitation, because they always regard them as prey. Keeping them in a brackish water aquarium is possible, and it would also be conceivable to set up a tank with an artificial tidal range. The main difficulty, however, is obtaining live specimens, as these are unlikely to be found in the delicatessen or aquarium trade.

Pelagic or Blue Swimming Crab, *Portunus pelagicus* (Linnaeus, 1758)

The **Pelagic Swimming Crab** is actually found in the Indo-Pacific and the Red Sea, but has inbetween also been recorded from the Baltic Sea. Therefore, it must be regarded as a displaced cosmopolitan in the foreseeable future, which has probably already been spread worldwide by ships. This Swimming Crab can reach a legspan of about 20 centimetres and is therefore also commercially fished and traded. It is then often declared as **Blue-Crab** or **Flower-Crab**; however, other similar species are also traded under this name, such as the **Crucifix Crab *Charybdis feriata*** (picture below), for example. The teeth of its claws alone reveal the strong predator, which feasts on anything it can overpower. It would be interesting to keep such animals in an aquarium, but then they would have to be kept alone in an appropriately large species tank. Unfortunately, due to lack of demand, such crabs are hardly ever imported alive, while frozen specimens can often be bought in Asian grocery shops. Although Swimming Crabs are basically marine species, this species was included in the canon of this work, because it also regularly migrates into estuaries and rivers and can temporarily survive in habitats with brackish water and very low salinity. In this respect, it could well happen, that small juveniles accidentally reach us as by-catches of pet fish consignments from tropical or subtropical coastal regions.

Green Shore Crab , *Carcinus maenas* (Linnaeus, 1758)

Juvenile Shore Crabs make prey of Blue Mussels.

Big male Shore Crab.

The common **Green Shore Crab** is probably the most common crab of the North Sea and the Baltic Sea. It reaches a carapace width of up to 8 centimetres; the females remain somewhat smaller. This species was included in this work, just like the following species, because it tolerates even very low salinities and can thus be kept in a brackish water aquarium at room temperature without any problems. In the Baltic Sea I found this species at a salinity of only 1.008 in brackish water. And furthermore they can also be found in harbours and estuaries. During the low tide they often sit in tide pools or under stones, where sometimes whole crab nests of up to 100 specimens squat on top of each other, waiting together for the next flood. They tolerate salt density fluctuations and other extremes such as heat and coldness without any problems. If they have not found cover under stones during low tide, they hide by simply burying themselves into the sand. They can also survive several days without water if they are stored in high humidity. Shore Crabs are very predatory and will eat anything they can overpower. The fact, that they sometimes also eat smaller or freshly moulted conspecifics makes them cannibals in the eyes of some authors. However, this is not their preferred way of feeding, because they are actually more interested in using the remains of other animals or also young mussels, worms and other things as source for their nutrition. Therefore, I would classify them as "secondary" cannibals, similar to the lobster or the pike. The Shore Crab plays the role of health police in the ocean and is also itself an important food source for seabirds and certain fish and crab-species. Shore Crabs can be picked up with the bare hand, preferably grasping them from behind by the outermost marginal spines of their head breast, while at the same time pressing them lightly against the ground in the middle so that they cannot break out sideways. This is because Shore Crabs prefer to run sideways, which earned them in Germany the popular name "Dwarslöper" among the coastal population. This is a special sensory-physiological performance of the crab's nervous system to coordinate all legs in alternating steps, so that the crab does not fall down. Experiments have even shown, that the Shore Crab is able to compensate for additional weights attached to one or more of its legs. Shore Crabs are easy to keep in the aquarium, but they are also mastering at "escaping", which is why the tank should be well covered. Since they eat everything, they can overpower, you have to think carefully about which animals you associate them with. If the Shore Crab is held down or cornered, it may also part with superfluous limbs, which break off at special predetermined breaking points. These points are recognisable as distinct indentations at the base of its legs. If a leg breaks off, the blood supply is immediately cut off, so that the crab does not bleed to death. Shore Crabs can easily regenerate lost body limbs according to the cactus sprout principle: At first, a grey gelatinous mass of tissue, which looks like a sack, oozes out of the stump of the base of the leg. After a while, this takes on the basic shape of the first leg segment in a slightly smaller version than the legs still present. Then another leg segment sprouts from this segment until the leg is formed so far as a small "bag leg". Then the crab moults and has a fully functional new leg in a smaller version on its body again. The males of the Shore Crabs can be recognised by the fact, that they only have a narrow "pocket" under their ***Cephalothorax***, while the females carry a wide “pocket” (the ***Pleon***). Days before Shore Crabs mate, the males collect a mating female, drag her around, clutching her belly and wait for the female to moult. Then the ***Copula**** occurs, whereby the male pushes his "pouch" under the female's, thus transferring his sperm package. The female then ejects her orange eggs and allows them to be fertilised by the sperm package. After a few weeks, these eggs hatch into small ***Zoea larvae***, which first swim freely in the plankton, then transform into ***Megalopa larvae*** and then into small crabs, which become bottom-dwelling. If you go foraging among kelp and under rocks, you can often spot the small pinhead-sized crabs. They also often live among the webs of mussels, where they search for small worms and other prey. When the young Shore Crabs have reached sizes of about 1centimetres carapace width, they can be found in all kinds of colour patterns, ranging from brown to green to white and orange patterned. These patterns camouflage them excellently among the barnacles and algae of the intertidal zone. In winter, especially the large Shore Crabs migrate to deeper water, as the mudflats then become too cold for them. It is not until spring, usually from April, that the larger Shore Crabs reappear in larger numbers on the mudflats. In the mudflats, however, Shore Crabs not only have to contend with the rigours of the seasons, but also with competition from other introduced crab-species. These include the **Chinese Mitten Crab** ***Eriocheir sinensis***, from the Talon-Crab family ***Grapsidae***, the **Japanese Shore Crab** ***Hemigrapsus penicillatus*** and the **Pacific Shore Crab** ***Hemigrapsus sanguineus***. On the other hand, the Shore Crab has already been spread worldwide by ships, so that it can now also be found in San Francisco harbour, for example, where it grows even larger than in the North Sea and can reach carapace widths of up to 10 centimetres. This ensures the preservation of the species in any case, even if it should become extinct on the German coasts due to displacement from the Far East. Thus, the Shore Crab is one of the clear winners of general globalisation and can already be classified as a modern cosmopolitan.

Mediterranean Green Crab, *Carcinus aestuarii* Nardo, 1847

This Shore Crab has no rostral spines between its eyes.

The **Mediterranean Green Crab** can be distinguished from common **Green Shore Crabs** essentially by a lack of frontal spines between its eyes. It is normally found around the Mediterranean Sea, although it can also be found in lagoons and estuaries. Their size is about the same as our native Shore Crab . Since this species also tolerates low salinity, it is as durable as the previously described North Sea and Baltic Sea Green Shore Crab . It is unclear, whether the status of the Mediterranean species can be maintained, or whether it is in fact only a subspecies. Furthermore it is quite possible, that both species are already very variable within the species or mate with each other. But it is a fact, that since 2009 I could find several animals in Wilhelmshaven, but also in Norddeich, which showed characteristics of the Mediterranean crab species. Especially the absence of the frontal spines should be mentioned. Unfortunately, it cannot be assumed that these were non-hybridized pure-bred specimens, which makes it difficult to identify them as a species by conventional means. Should the specimens I found really be ***Carcinus aestuarii***, this would be further evidence of the progressive climate change in the southern North Sea. A warning sign from mother nature to our address, which better should not be ignored.

Infraorder *Anomura* – Squat Lobsters, Porcelain Crabs & Hermit Crabs

The **Infraorder *Anomura*** comprises very different families, such as **Hermit Crabs, Squat Lobsters** and **King Crabs**. What they all have in common is, that at least the last of their five pairs of legs is severely atrophied, and usually folded or hidden under the ***Pleon***. In the Hermit Crabs, even the last two pairs of legs are reduced to small stumps. In addition, the ***Abdomen*** of the Hermit Crabs is soft and unarmoured as well as turned to the right, so that it can be housed in a snail shell without any problems. The smaller species of Hermit Crabs are not of interest to commercial fisheries and are therefore not threatened in their existence. Some species, such as some Hermit Crabs, benefit from fishing because they use dead bycatch animals as food. However, the introduction of species, such as the **King Crab *Paralithodes camtschaticus*** from the North Pacific to the northern North Sea, has impacts on the ecosystem that are not yet fully foreseeable. Although King Crabs are already commercially fished and successfully marketed, this fishing is not sufficient to control the invasion. Fortunately, these animals are distinctly cold-water dwellers, so that they can hardly spread further south. If general climate warming continues, this could one day mean the end of these cold-loving animals in the North Sea, as they only feel comfortable and reproduce at low temperatures. Furthermore, marine ***Anomura*** generally do not tolerate low salinity levels in the long term, which is why they do not occur in the southern Baltic Sea. Only very few species of the **Order *Anomura*** live regularly in freshwater, such as the species of the genus ***Aegla***, which belong to the immediate family of the marine **Squat Lobsters (*Galatheidae*)**. However, also of interest to freshwater aquarists are the **Land Hermit Crabs (*Coenobitidae*)**, which carry out their larval life cycle in the sea, but the rest of their life on land. Strictly speaking, they belong in a terrarium rather than an aquaterrarium. But it would be perfectly reasonable to construct a tropical brackish water habitat with mangrove roots and tidal influence, into which one could also introduce marine Hermit Crabs, which can tolerate low salinities. Because King Crabs and Porcelain Crabs on the one hand and Hermit Crabs on the other show very strong morphological and anatomical differences, I classified them as ***Anomura I*** and ***Anomura II*** in the systematics presented at the beginning. This is because I would consider it quite possible, that this part of the taxonomy will be revised in the foreseeable future and possibly further taxa will be created here. Therefore, statements on the systematic position of some animals can unfortunately always be assumed to be provisional and subject to reservation. I therefore recommend, that the currently valid classification be perceived only as a suggestion and not as objective truth.

Family *Aeglidae* – Freshwater Squat Lobsters

These species of **Freshwater Squat Lobsters**, which are often referred to as **"Crabs"** or **"Cockroach Crayfish"** in aquaristics, are only found in a very small number of freshwater habitats worldwide. Several species are known from South America, where they are distributed from Argentina to Brazil. They all belong to the genus ***Aegla***. Unfortunately, they are rarely imported, as they are usually unattractive in colour and shy. Nevertheless, they are interesting to keep, as their behaviour is still largely unexplored and as they can even be reproduced without any problems. However, they are not as productive in breeding as shrimps and crayfish.

De La Plata Squat Lobster, *Aegla platensis* Schmitt, 1942

The **De La Plata Squat Lobster** is found in the rivers of Argentina, Brazil, Paraguay and Uruguay. It reaches a carapace length of about three centimetres. It is typical for this species that the last pair of legs is strongly atrophied and carried hidden under the ***Pleon***, so that it cannot be seen from above. The species keeps well in the aquarium, but should only be kept at room temperature according to its mostly moderately warm origin. The aquarium must have many hiding places and should be well ventilated, especially in summer, as the animals come from habitats with a lot of current. The species reproduces moderately well in the aquarium depending on local water values, but not every breeder is lucky enough to be successful with them. A socialisation with fish is self-explanatory, as the freshwater squat lobsters otherwise either remain extremely shy or hunt and feed on the fish also.

Family *Coenobitidae* – Land Hermit Crabs

Land Hermit Crabs are found all over the tropical seashores and there are several species, which are not easy to distinguish. Land Hermitsspend their lives in two different cycles, living on land as adults but having to handle the larval and juvenile stages in the sea. This means, that the adult Land Hermitsalways stay close to the coast, and therefore hardly ever venture inland. The species of the genus ***Coenobita*** carry snail shells around with them, as all marine Hermit Crabs do, while the **Palm Crab** of the genus ***Birgus*** (see illustration centre right) as an adult no longer needs a snail shell, but folds its ***Abdomen*** under its body. The **Palm Crab *Birgus latro*** is also the largest land-dwelling crustacean of all, as it can reach a wingspan of more than 40 centimetres and weigh up to five kilograms. This is why it is also eaten locally, but must then be kept separate for a few weeks beforehand, as it could otherwise have eaten poisonous plants, making it inedible for human consumption. Land Hermits feed on fruits and carrion, and the Palm Crab also eats coconuts, which it can actually pick off the palms itself. The Palm Crab has strong claws, with which it can crack coconuts, so it must be handled with care. It is also known that these animals can even destroy wire casings of cages! But also the small **Land Hermits** of the genus ***Coenobita*** can pinch sensitively, if you touch them carelessly. They are occasionally imported for terraristics, and red, purple or brown terrestrial hermits are commonly offered in the trade. You should offer them a terrarium with a water bowl, plus climbing opportunities and a cuttlebone for the calcification of their shell. It is possible to keep several Land Hermits in one tank at the same time, but then there should always be enough space and a sufficient number of snail shells to avoid quarrels. The Palm Crab can also be kept in a terrarium, but a container for such a big animal must be appropriately large.

Indonesian Hermit Crab, *Coenobita brevimanus* Dana, 1852

The **"Indonesian" Hermit Crab** is also native to Madagascar, the Seychelles and Tanzania. It reaches a carapace length of about five to six centimetres. It is best kept in a terrarium with potting soil or sandy soil. In order to grow, Land Hermitsalso have to moult like other crustaceans. They usually burrow into the substrate and disappear there for a few days. Since they grow by moulting, you must of course offer them a sufficient number of empty snail shells. In the acclimatisation phase they can be very shy, so that you hardly see them at first. They then retreat into the sulky corner of their shell, which they close completely with their legs and large scissors. Under no circumstances should you pull them out of their enclosure, as this is more likely to tear off their ***Abdomen*** than to cause them to leave their enclosure. This fact should be made clear to children in particular before acquiring such pets. Children should also be made aware, that terrestrial hermits can be very sensitive to pinching if they are not feeling well. Since terrestrial hermits are robust and relatively low-maintenance terrarium animals, attempts have been made in the past to market them as special pets for children. However, this should always be viewed somewhat more critically and accompanied by responsible parents to prevent the loss of the animal or animal cruelty. Furthermore, it should always be remembered, that Land Hermits often have a predominantly nocturnal lifestyle, so that they should not be constantly disturbed by their owners during the day.

Class *Merostomata* – Horseshoe Crabs („Feeler-Less")

" Horseshoe Crabs" belong systematically to the **Class *Merostomata***, the "feeler-less". And not only the lack of antennae, but also their special anatomy distinguishes them as something own. Which is why taxonomists do not include them among the crustaceans, but have placed them close to the arachnids (spiders and scorpions). On the top of their head they have 5 eyes, which puts them closer to spiders, and the division of their body into three parts places them close to the extinct Trilobites and the recent Slaters. These animals are living proof, that it is not always easy to draw a clear dividing line between freshwater and saltwater aquaristics, as in their natural ranges they may well enter mangrove areas and estuaries, where brackish water with low salinity is present. This is why they are sometimes offered for freshwater aquaria. However, they are basically sea dwellers, which also occur in the mangrove zones of Southeast Asia with the genus ***Carcinoscorpius***. Juveniles of the **Caribbean Horseshoe Crab *Limulus polyphemus*** are commonly offered for aquaristics. Numerous fossil specimens of Horseshoe Crabs from the genus ***Mesolimulus*** have been found in the slab limestones of the Bavarian village of Solnhofen. In past geological eras, there were even species that reached a total length of about 180 centimetres!

Horseshoe Crab, *Limulus polyphemus* (Linnaeus, 1758)

Somehow it looks like a flower pot with legs at the bottom...

Free-swimming Horseshoe Crabs swim on their backs.

The **Atlantic Horseshoe Crab** ***Limulus polyphemus*** occurs on the west side of the Atlantic Ocean from the Yucatan Peninsula in the south to Maine in the north. Isolated specimens have also been caught in the North Sea, because they were probably released here. Horse shoe crabs may enter estuaries and brackish water under natural conditions. This is why this interesting animals have been integrated in this work. When fully grown, Horseshoe Crabs can reach a total length of about 60 centimetres. For most of the year, the animals live at depths of up to 200 metres on the continental shelf, but during their spawning season they even come as far as the shallow water area on the shore. Juveniles regularly come into the pet trade and are offered there for keeping in marine aquaria. As their body armour is thin and fragile, they should not be kept with predatory crustaceans or fish. They should also not be kept warm permanently, as in their natural habitat they are only ever exposed to heat influences seasonally. They feed on all kinds of invertebrates that they find in the substrate, especially worms! For this reason, one often comes across the advice in the literature to feed the animals with *Tubifex-worms* by hand. The animals can be taken out of the water and turned onto their backs. When horse shoe crabs swim, they prefer to turn on their backs and move forward using the pulsating leaf-like gill appendages under the 2nd body segment. They can, of course, also walk on their legs across the bottom, which then looks like a flowerpot taking a walk. Unfortunately, you don't see them very often, because they like to burrow into the substrate for days. You can use fine shell grit as substrate, in which numerous bristle worms, amphipods, isopods and other small animals cavort. Fine sandy substrate is also suitable. Since a tropical sea aquarium is too warm for these animals in the long run, they must be relocated to an unheated seawater tank for permanent keeping. As these animals can reach a final size of 60 centimetres, they are not suitable for permanent care in a reef aquarium. However, they are very interesting to keep and their peculiar locomotion and behaviour are always very fascinating.

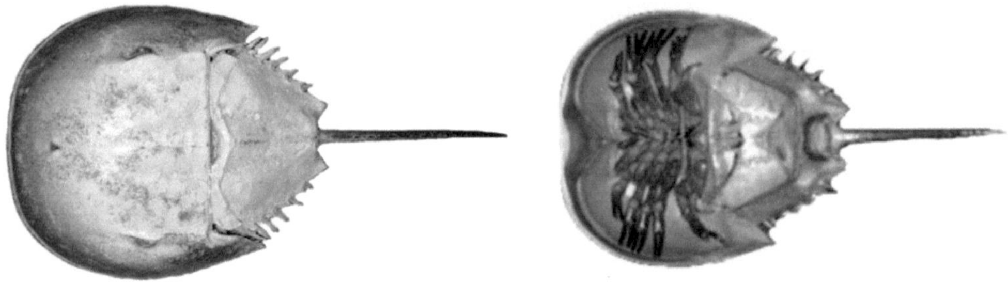

Horseshoe Crab *Limulus polyphemus* from above and below (preserved specimen).

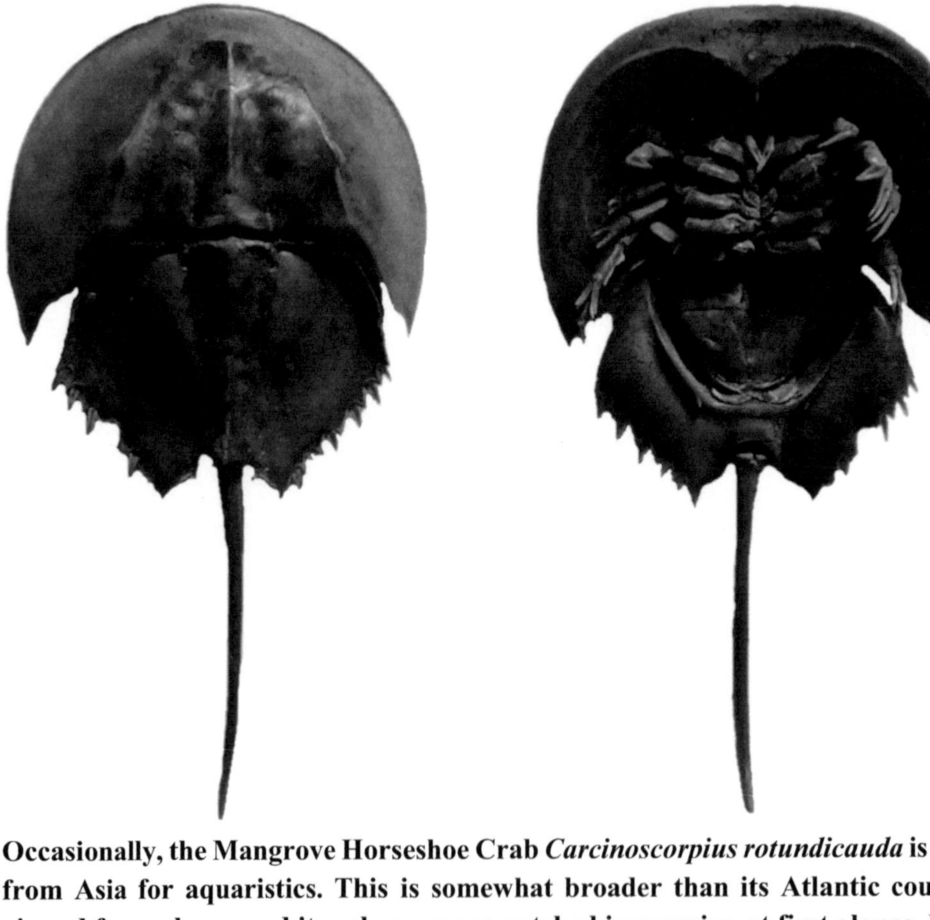

Occasionally, the Mangrove Horseshoe Crab *Carcinoscorpius rotundicauda* is imported from Asia for aquaristics. This is somewhat broader than its Atlantic cousin when viewed from above, and it makes a more notched impression at first glance. However, in the interest of the declining populations of this species in many areas, it would be well to refrain from acquiring it. This species is also known to occasionally migrate into brackish water. For permanent care of adults, however, one will not be able to avoid setting up an appropriate seawater aquarium.

An Aquarium for crayfish and shrimps

The 80-centimetre aquarium for Apricot Crayfish, *Neocaridina*-Shrimps and Dwarf Neon Rainbowfish.

The aquarium measures 80 X 46.5 X 35 centimetres (width times height times depth). This gives it a gross capacity of approx. 130 litres.

The substrate

The substrate consists of fine sand with a grain size of 0.1-0.4 millimetres. The layer height is about 5 centimetres. In addition, the sand in this aquarium has been enriched with a few zeolite stones, which not only have an additional denitrifying effect, but also serve as occupational therapy for the crayfish.

The lighting

The tank is illuminated with a 12-watt aquarium LED in the colour cold white, which is about 50 centimetres long. For plant growth and optimal illumination of the tank, this one tube is completely sufficient. A glare shield has been mounted in front of it. As the crayfish kept here are cave-dwellers in nature, the lighting must not be too bright anyway.

Filtration

The water is filtered through a JBL Cristal Profi Green Line E702 external filter. For tanks from 60 to 200 litres. The filter pot contains foam, cotton wool, zeolite stones and some activated carbon as filter media. The carbon is supposed to counteract the yellowish tinge of the water caused by the wood. The main purpose of the external filter is to ensure continuous water circulation. Every time I clean the filter, I sieve the detritus with a net, because unfortunately small shrimps keep getting into the filter, where they seem to thrive.

The background (the back wall) and the planting

A multi-part polyurethane aquarium back wall was attached to the background pane, on which some **Java Moss** of the species ***Taxiphyllum barbieri*****, (synonymous with** ***Vesicularia dubyana*** **and** ***Hypnum dubyanum*****)** was attached, which has grown there in the meantime. The Java moss is also growing on the wood that was brought in. In addition, some free-floating **Hornwort (*****Ceratophyllum demersum*****)** was brought in, which is to remain hanging between the root branches for purely decorative reasons. It is not planted in order to reinforce the impression of a random plant driven by the current.

The root woods

The wood brought in is two roots that have simply been turned upside down and are actually lying upside down in the basin. The optical effect is therefore based on the compositions of the Japanese aquarium designer Takashi Amano, who sadly passed away in 2015. The wood is pieces that I received from a crab cutter in the North Sea. They have therefore been leached in salt water and thus preserved in the best possible way for aquaristics. Before I put the wood into the tank, however, I first had to free it from the barnacles and other growth that had grown on it. Such pieces of wood are regularly torn out of the shore fortifications near the coast by storms. Often, such loose wood can also be found near dikes, which are located in front of the dikes.

Microfilms of algae and bacteria form on the beech leaves, which are an important food source for both snails and shrimps. The infusoria living here are also important for the broods of crayfish and shrimps.

A juvenile Apricot Crayfish explores the back wall. Many structures in the aquarium provide movement and occupation for the animals and thus also serve to reduce stress and prevent aggression.

The Hornwort catches itself between the root wood. The Dwarf Neon Rainbowfish provide a nice contrast.

Shrimps in Java Moss feed on a food tablet.

The water

Our local tap water is rather soft and contains no hardness builders, nitrates or other pollutants, which is of great advantage for keeping freshwater shrimps. The current water values with Sera's standard quick test look something like this: Nitrate: 10mg; nitrite not detectable; total hardness below 3; carbonate hardness below 6; pH value 6.8 and chlorine is not detectable. But the crayfish also like this environment and show their best colours here. No additives were used to make the water a little harder. The tank is kept at room temperature without heating. This means that the water temperature can vary between 20° - 22° Celsius in winter and 25° - 27° Celsius in summer. When a water change is carried out, usually about 80 percent of the water is changed, whereby the change water is appropriately tempered and pumped directly from the tap into the tank without further treatment.

The stocking

The tank is stocked with a total of 4 specimens of the **Apricot Crayfish *Cherax holthuisi***, the largest of which is the male father. The other three crayfish are all descendants of this male from their own breeding; unfortunately, the female has already passed away. However, there is at least one female among the offspring, so the crayfish continue to reproduce. However, I am not keen on breeding large numbers of crayfish, which is why I have also refrained from using perforated bricks or special crayfish burrows. However, the animals dig their own tunnels under the wood or even climb around in the hornwort. This shows me that they use all available structures. This avoids aggression among the crayfish, as they are "busy". And I have never been able to observe any attacks against conspecifics, even during moulting. Aggressive animals in the aquarium document serious husbandry errors, which are usually due to a lack of structure within the respective aquarium system. This is a rule that, in my estimation, applies to almost all species that can be kept in an aquarium. Among other things, the structures implanted in the tank must be such that the animals kept cannot see each other all the time; this rule applies to cichlids just as much as to any crayfish or other crustaceans. Personally, I consider the Apricot Crayfish to be the best possible aquarium crayfish of all, as it belongs to the troglodyte species from Indonesia. This means that this species lives in caves in nature, where it feeds on foliage and detritus. Conversely, this means that these crayfish do not feast on aquarium plants, as these are not found in their natural habitats. If an apricot crayfish digs up a plant, it does so only to find a way through. However, it will never eat it! Furthermore, there are now more than 100 freshwater shrimps living in this aquarium, a good part of which belong to the ***Neocaridina davidi*** complex. Some were originally traded as **"Sakura"**, other specimens have dorsal stripes and other not so spectacular colour characteristics. Some of them were introduced with the hornwort from another aquarium. This ensures at least a certain genetic diversity, which prevents permanent inbreeding. And since I have no interest in commercial breeding, there is no need to keep any strains "pure" here. In addition, I had put some **Ramshorn Snails (*Planorbarius spec.*)** in the tank. However, these were not meant to rid the panes of algae, but to serve as food for the crayfish. This did not work, however, as the crayfish initially showed no interest in this prey, which I then had to collect and reduce by hand myself. In the meantime, however, the snail population has decreased noticeably, so I wonder if the juvenile ***Cherax holthuisi*** have reduced it. However, I could never observe an attack on a snail! Finally, I should mention that there is a single **Dwarf Neon Rainbowfish (*Melanotaenia praecox*)** swimming patrol in the tank, the last of the original four specimens. It does not prey on the adult shrimps, which can be seen from the fact that they swim freely through the water, especially in the evening hours. I suspect, however, that he grabs a few juvenile shrimps now and then, thus regulating their total population somewhat.

Feeding

The animals are fed with regular frozen, flake and granulated food, whereby I do not prefer any special types. I also add some dry beech leaves to the tank from time to time, without having treated or watered them beforehand. After a day it sinks by itself and is then used by the shrimps and crayfish alike.

The Wall - Or: The Great Wall

Apparently, my crayfish have worked together to build a 30 cm long wall of small stones across the tank under the roots. Whether the wall serves as a demarcation of territories seems to be questionable, as the 4 crayfish harmonise very well with each other and I have never observed any scuffles or attacks against conspecifics. Not even during moulting. Perhaps the structure is simply to create hiding places for the fry.

Here, an adolescent Apricot Crayfish participates in building the wall. It is remarkable that this wall is demonstrably the work of several individuals.

Moulting

The large male recently moulted directly in front of the front window, so I was able to document the final part of this moult with my mobile phone camera. The most striking thing was that the animal showed a bright deep orange before moulting and somewhat paler colours afterwards. I collected the ***Exuvium*** and examined it. I then noticed, that this male has both: Male and female gonopores. Should this specimen bear brood one day, this would be proof, that the **Apricot Crayfish *Cherax holthuisi*** is also a true hermaphrodite.

Apricot Crayfish shortly after moulting; the intense colouration has now faded somewhat.

Construction plan of a decapod

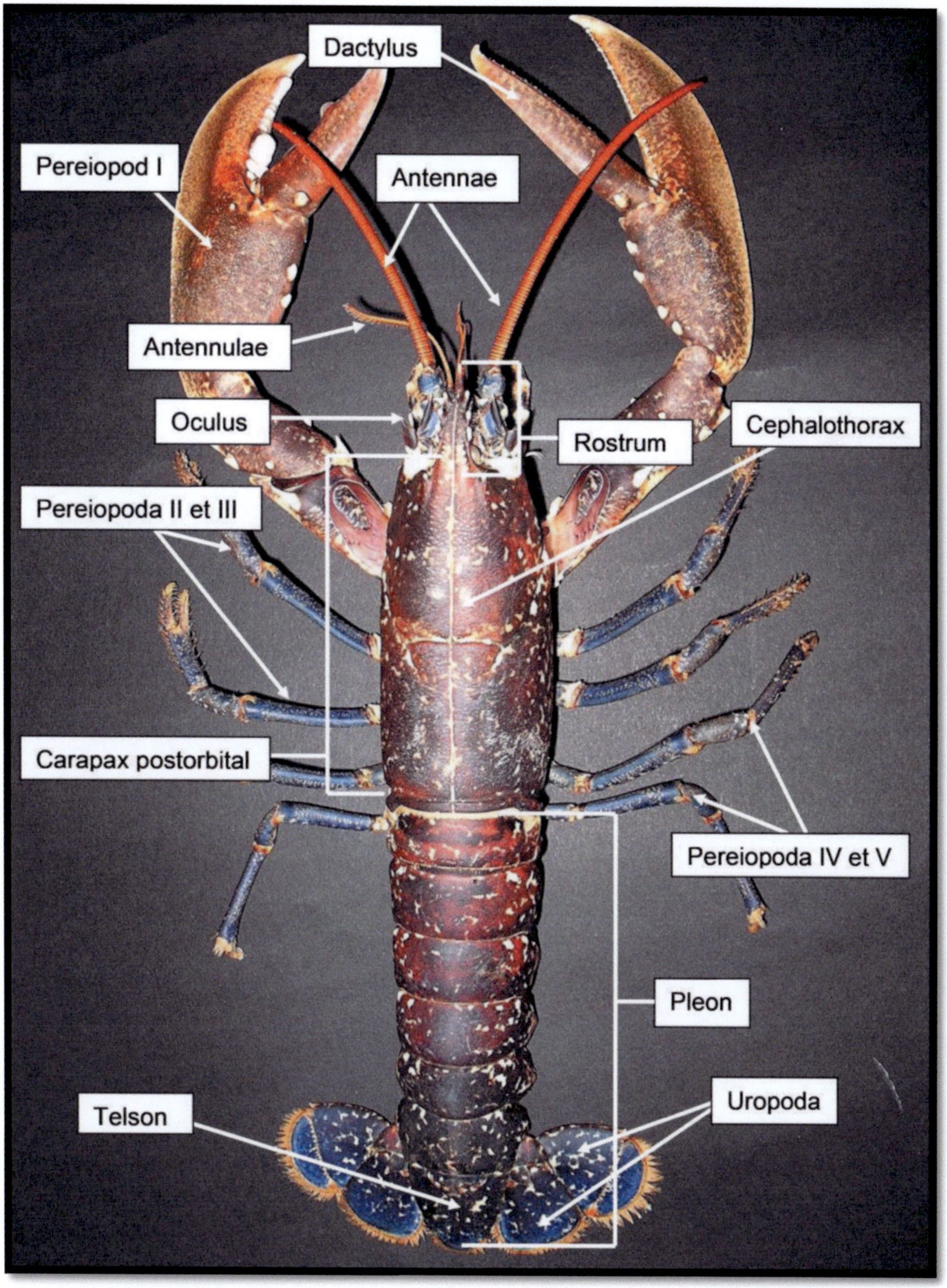

***Dactylus* = claw-finger; *Pereiopoda* = legs; *Pleon* = tail; *Uropoda* = tail fans.**

Construction plan of a female decapod from below

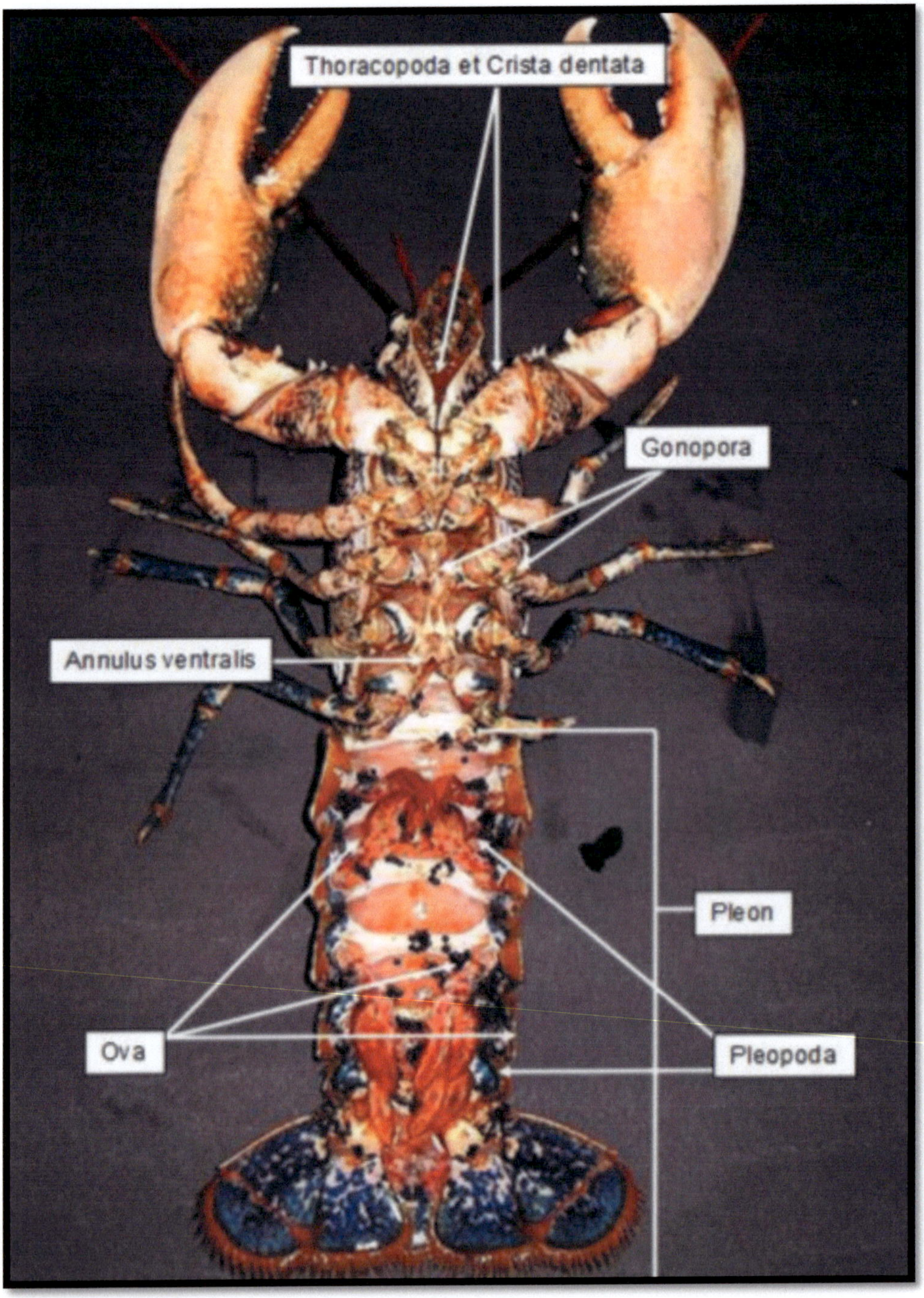

Interior of a decapod

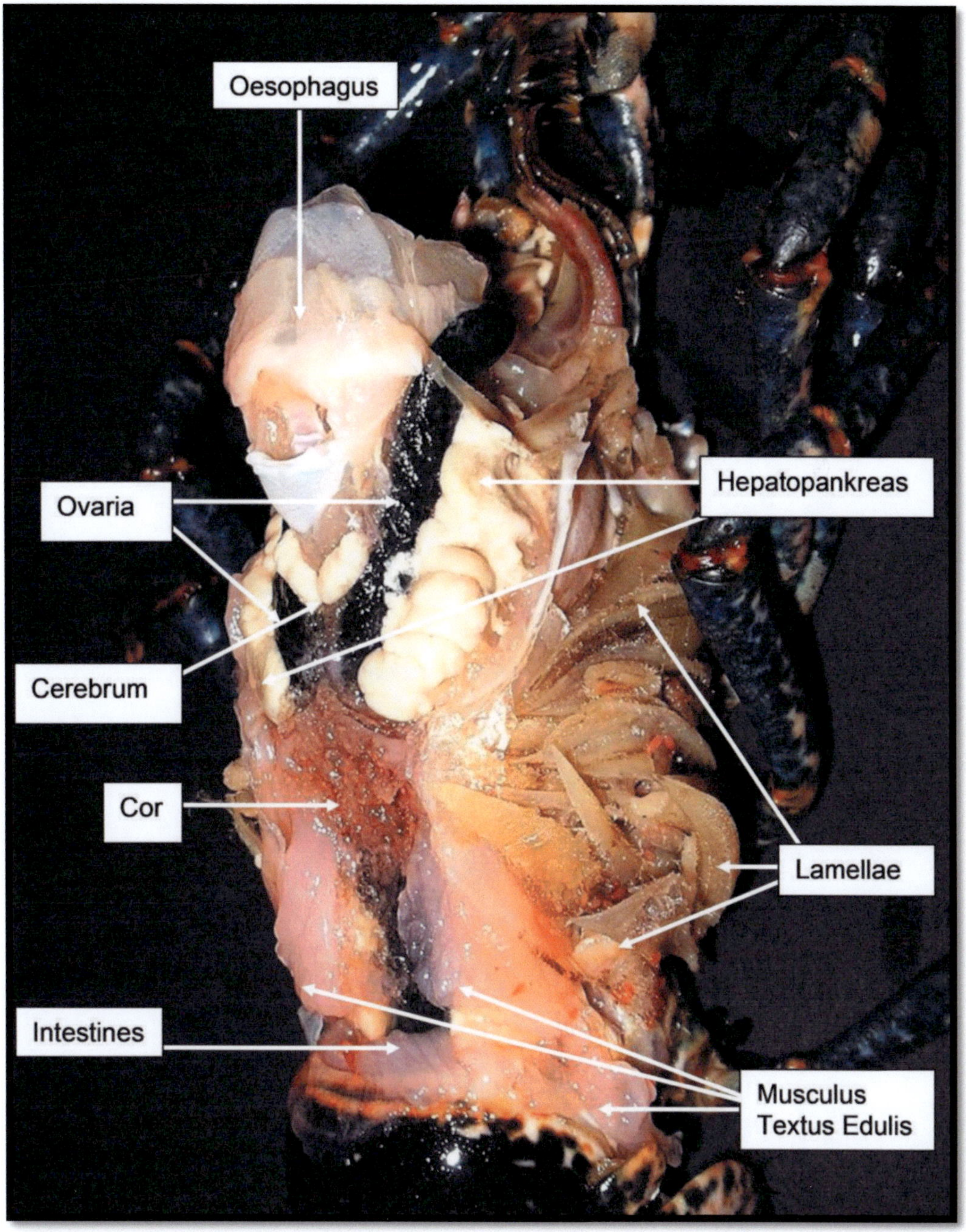

Oesophagus **= stomach;** ***Hepatopankreas*** **= midgut gland;** ***Ovaria*** **= ovaries;** ***Cerebrum*** **= brain;** ***Cor*** **= heart;** ***Lamellae*** **= gills;** ***Intestines*** **= gut;** ***Musculus Textus Edulis*** **= edible meat.**

Interior and construction plan of a crab

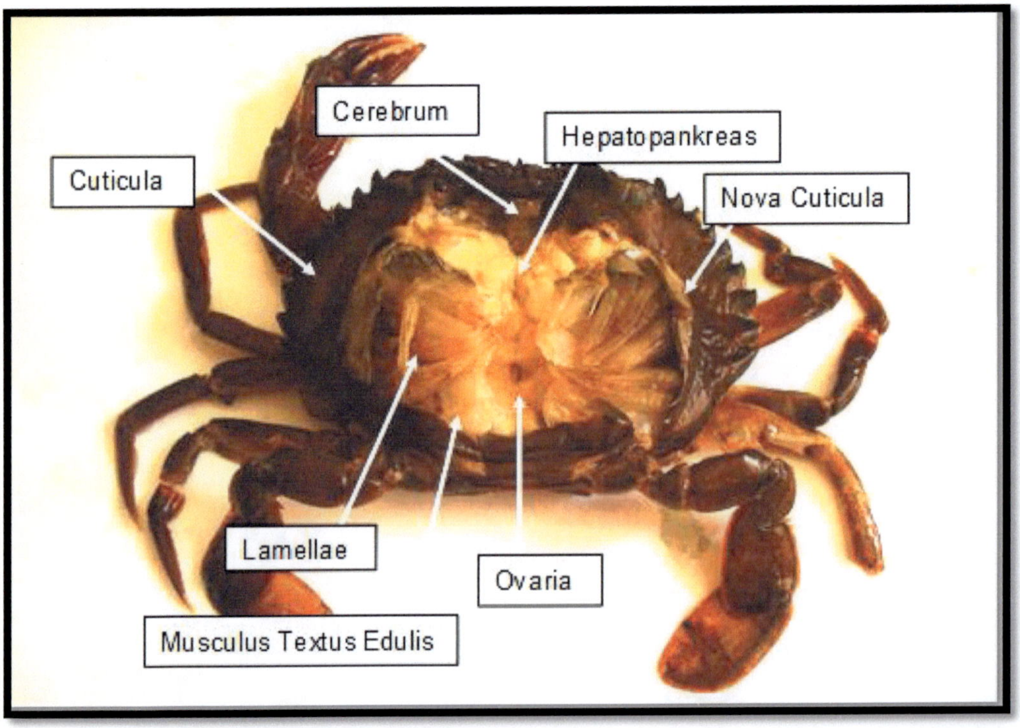

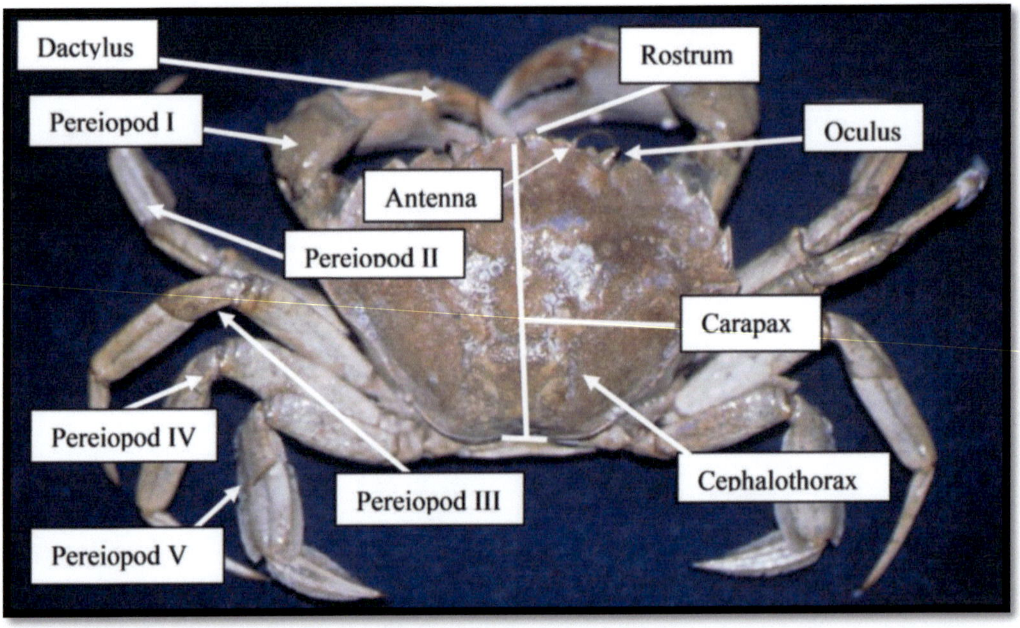

Bibliography and sources

Books:

Karl-Heinz Zeitler: Muscheln, Krebse, Schnecken, VFSF 1990, ISBN 3-490-29514-5.
Helmut Debelius: Krebsführer, Weltweit, I. Auflage 2000,
ISBN 3-86132-504-7.
Wilhelm Eigener, Enzyklopädie der Tiere, Georg Westermann Verlag, 1979,
ISBN 3-14-508000-8.
Chris Lukhaup: Süßwasserkrebse aus aller Welt, Dähne Verlag, 2003,
ISBN 3-935175-14-0.
Chris Lukhaup/Reinhard Pekny: Krebse im Aquarium, Dähne Verlag, 2005,
ISBN 3-935175-31-0.
Andreas Karge/Werner Klotz: Süßwassergarnelen aus aller Welt, Dähne Verlag 2007,
ISBN 978-3935175-39-5.
Peter Hunnam, Lebensraum Aquarium, Bechtermünz Verlag, 2002,
ISBN 3-86047-416-2.
Blanke, Dietrich; Flusskrebse (*Astacidae*) in Niedersachsen; Information des Naturschutzes Niedersachsen, 18. Jg., Nr.6, 146-174, Hildesheim 1998
Ludwig Grieser, Meeresfrüchte, Falken Verlag. ISBN: 3-8068-0886-4.

Internet:

http://www.wikipedia.org/
http://www.wirbellose.de/
http://www.crusta10.de/templates/index.php?lang_id=1&showid=19&katid=1
http://www.edelkrebs-niedersachsen.de/
http://iz.carnegiemnh.org/crayfish/NewAstacidea/infraorder.asp?io=Astacidea
http://zipcodezoo.com

Thanks!

I would like to thank all my friends and acquaintances who gave me valuable tips for writing this little book. In particular, however, I would like to thank my family, who let me write this little book. Another thank you goes to Steffen Göckemeyer, who provided me with real Noble Crayfish, and to all the many hobbyists with whom I have exchanged experiences and sometimes animals over the past years. I am also grateful to dealers, importers and various crayfish breeders, who have always given me insight into the handling of the animals. Without this rich exchange of information, this book would never have come about in this form.

Glossary

Abdomen	The tail of crustaceans; also called ***Pleon***.
Adult animals	Grown and reproductively mature animals.
Species concept in biology	Often controversially discussed and understood; probably there is still no definitive definition here in biology.
Estuaries	River mouths in the sea.
Autochthonous species	Species that only occur in very specific habitats and cannot exist without them.
Benthic	Ground-oriented.
Carcinologist	Crustacean researcher.
Cephalothorax	Head breast of the crustaceans.
Chitin	Polysaccharide from which crustaceans & insects form their shells.
Dactylus	Movable scissor finger of the crustaceans.
Decapoda	Crustaceans with ten feet.
Endemic Species	Species that are only found in a specific location.
Exoskeleton	Outer skeleton.
Exuvium	Stripped empty skin of a crustacean after moulting.
Juveniles	Offspring.
Larvae	The smallest juvenile animals, which do not yet resemble the parents in appearance, but which still have to develop further through a transformation known as metamorphosis before they resemble the adults.
Megalopa	Larval stage of various crab species.
Molecular genetics	Branch of biology that deals with the genetic material of animals and plants in order to clarify species concepts and identify species.
Morph	Outer appearance of a species.
Ecology	The study of the relationships between organisms.
Osmosis	Unidirectional diffusion of a substance through a semi-permeable membrane, "vapour pressure gradient".
Pereiopoda	Legs of Crustaceans.
Pleon	Tail of Crustaceans; ***Abdomen***.
Pleopoda	Swimming feet below the ***Pleon***.
Rostrum	Crustacean's spiny thorns between the eyes.
Taxonomy	Branch of biology responsible for relating animals and plants to each other according to degrees of relationship.
Telson	The middle part of the tail fan of a decapod crustacean.
Uropoda	The appendages of the crustaceans on the tail fan.
Zoëa	Larval stage of many crustaceans.

Index of Latin nomenclature

About the author

The author Sven Gehrmann, born in 1969, now sees himself as a cryptographer, because in his publications he always struggles to make the invisible visible. He was born in Berlin and is now currently living in Norden near Norddeich on the coast of Lower Saxony. He has been interested in everything that lives under water since he was a child. He has always been particularly interested in and fascinated by crustaceans and fish. Since 1983, he has been an enthusiastic hobby aquarist and nature fan of our native aquatic animals, especially North Sea animals. In his cellar he keeps a collection of various preserved species. So far, he has published various articles in aquarium journals, as well as an eco-thriller, several non-fiction books about marine animals and some poetry books.

You can find him on the internet at: **WWW.NORDSEEFAUNA.ORG**.

In his publications, he never minces his words and calls a spade a spade, since obviously no one else does. In doing so, he takes no account whatsoever of the false kind of "political correctness" that has been successfully installed everywhere here in order to preserve the appearance of decency. Nor does he profess any allegiance to any political party or direction, but only feels committed to the cause of a dying nature, that seems to escape public perception in large parts.

Bibliographic information of the German National Library:
The German National Library lists this publication in the German National Bibliography; detailed bibliographic data are available on the Internet at http://dnb.d-nb.de.
Imprint: GENERAL FRESHWATER-CRUSTACEAN PRACTICE
Fresh- & Brackish-Water Crayfish, Shrimps, Prawns, Crabs, Hermits & Horseshoe Crabs

Production and publisher: BoD - Books on Demand, Norderstedt.
ISBN: 9783753497983.

9 783753 497983